Forecasting in Mathematics - Recent Advances, New Perspectives and Applications

Edited by Abdo Abou Jaoude

Published in London, United Kingdom

IntechOpen

Supporting open minds since 2005

Forecasting in Mathematics - Recent Advances, New Perspectives and Applications
http://dx.doi.org/10.5772/intechopen.87892
Edited by Abdo Abou Jaoude

Contributors
Zineb Aman, Latifa Ezzine, Younes Fakhradine El Bahi, Haj El Moussami, Yassine Erraoui, Isa Salman Qamber, Mohamed Al-Hamad, Sumit Saroha, S. K. Aggarwal, Preeti Rana, Deneshkumar Venegopal, Senthamarai Kannan Kaliyaperumal, Sonai Muthu Niraikulathan, Ismit Mado, Abdo Abou Jaoude, Hamza Turabieh, Alaa Sheta, Elvira Kovač-Andrić, Malik Braik

First published in London, United Kingdom, 2021 by IntechOpen
IntechOpen is the global imprint of INTECHOPEN LIMITED, registered in England and Wales, registration number: 11086078, 5 Princes Gate Court, London, SW7 2QJ, United Kingdom
Printed in Croatia

British Library Cataloguing-in-Publication Data
A catalogue record for this book is available from the British Library

Additional hard and PDF copies can be obtained from orders@intechopen.com

Forecasting in Mathematics - Recent Advances, New Perspectives and Applications
Edited by Abdo Abou Jaoude
p. cm.
Print ISBN 978-1-83880-825-9
Online ISBN 978-1-83880-827-3
eBook (PDF) ISBN 978-1-83880-828-0

For EU product safety concerns:
IN TECH d.o.o., Prolaz Marije Krucifikse Kozulić 3, 51000 Rijeka, Croatia, info@intechopen.com or visit our website at intechopen.com.

Meet the editor

Abdo Abou Jaoudé has been teaching for many years and has a passion for researching and teaching mathematics. He is currently Associate Professor of Mathematics and Statistics at Notre Dame University-Louaizé (NDU), Lebanon. He holds a BSc and an MSc in Computer Science from NDU, and three PhDs in Applied Mathematics, Computer Science, and Applied Statistics and Probability, all completed at Bircham International University through a distance learning program. He also holds two PhDs in Mathematics and Prognostics from Lebanese University, Lebanon, and Aix-Marseille University, France. Dr. Abou Jaoudé's broad research interests are in the field of applied mathematics, and he has published twenty-three international journal articles and six contributions to conference proceedings, in addition to three books on prognostics, applied mathematics, and computer science.

Contents

Preface

This book is titled Forecasting in Mathematics – Recent Advances, New Perspectives and Applications. Additionally, each time I work in the field of mathematical probability and statistics, I have the pleasure of tackling the knowledge, the theorems, the proofs, and the applications of the theory. In fact, each problem is like a riddle to be solved, a conquest to be won, and I am relieved and extremely happy when I find the solution. This proves two important facts: firstly, the power of mathematics and its models to deal with such problems and secondly the power of the human mind that is able to understand such problems and to tame such a wild concept that is randomness, probability, stochasticity, uncertainty, chaos, chance, and nondeterminism.

Mathematical probability and statistics are attractive, thriving, and respectable parts of mathematics. Some mathematicians and philosophers of science say that they are the gateway to mathematics' deepest mysteries. Moreover, mathematical probability and statistics denote an accumulation of mathematical discussions connected with the efforts to most efficiently collect and use numerical data subject to random or deterministic variations. In the twentieth century and the present time, the concept of probability and mathematical statistics has become one of the fundamental notions of modern science and philosophies of nature. This was accomplished after a long history of efforts by prominent and distinguished mathematicians and philosophers like the famous French Blaise Pascal and Pierre de Fermat, the Dutch Christiaan Huyghens, the Swiss Jakob Bernoulli, the German Carl Friedrich Gauss, the French Siméon-Denis Poisson, the English Thomas Bayes, the French Joseph Louis Lagrange and Pierre-Simon de Laplace, the English Karl Pearson and Ronald Aylmer Fisher, the Russian Andrey Nikolaevich Kolmogorov, the American John von Neumann, etc...

As a matter of fact, each time I read or meditate on these outstanding giants, I feel the respect, the admiration, and the esteem towards these magnificent men and giants of science who most of them were mathematicians, physicists, astronomers, statisticians, philosophers, etc... at the same time. They were, as we call them today: Universalists.

Moreover, this book develops methods for simulating simple or complicated processes or phenomena. If the computer can be made to imitate an experiment or a process, then by repeating the computer simulation with different data, we can draw statistical conclusions. Thus, a simulation of a spectrum of mathematical processes on computers was done. The result and accuracy of all the algorithms are truly amazing and delightful; hence, this confirms two complementary accomplishments: first the triumphs of the theoretical calculations already established using different theorems and second the power and success of modern computers to verify them.

To conclude, due to its universality, mathematics is the most positive and certain branch of science. It has been successfully called by philosophers the Esperanto of all sciences since it is the common, the logical, and the exact language of understanding, capable of expressing accurately all scientific endeavors. Although probability and statistics are approximate sciences that deal with rough guesses, hypotheses tests, estimated computations, expected calculations, and uncertain results, they still keep in them the spirit of "exact" sciences through their numbers, proofs, figures, and graphs, since they remain a branch of mathematics. Surely, the pleasure of working and doing mathematics is everlasting. I hope that the reader will benefit from it and share the pleasure of examining the present book.

Sincerely, I am truly astonished by the power of probability and statistics to deal with deterministic or random data and phenomena, and this feeling and impression has never left me from the first time I was introduced to this branch of science and mathematics. I hope that in the present book I will convey and share this feeling with the reader. I hope also that they will discover and learn about the concepts and applications of probability and statistics paradigm.

Abdo Abou Jaoudé, Ph.D.
Notre Dame University-Louaizé,
Zouk Mosbeh, Lebanon

Chapter 1

The Monte Carlo Techniques and the Complex Probability Paradigm

Abdo Abou Jaoude

Abstract

The concept of mathematical probability was established in 1933 by Andrey Nikolaevich Kolmogorov by defining a system of five axioms. This system can be enhanced to encompass the imaginary numbers set after the addition of three novel axioms. As a result, any random experiment can be executed in the complex probabilities set $\mathcal{C}$ which is the sum of the real probabilities set $\mathcal{R}$ and the imaginary probabilities set $\mathcal{M}$. We aim here to incorporate supplementary imaginary dimensions to the random experiment occurring in the "real" laboratory in $\mathcal{R}$ and therefore to compute all the probabilities in the sets $\mathcal{R}$, $\mathcal{M}$, and $\mathcal{C}$. Accordingly, the probability in the whole set $\mathcal{C} = \mathcal{R} + \mathcal{M}$ is constantly equivalent to one independently of the distribution of the input random variable in $\mathcal{R}$, and subsequently the output of the stochastic experiment in $\mathcal{R}$ can be determined absolutely in $\mathcal{C}$. This is the consequence of the fact that the probability in $\mathcal{C}$ is computed after the subtraction of the chaotic factor from the degree of our knowledge of the nondeterministic experiment. We will apply this innovative paradigm to the well-known Monte Carlo techniques and to their random algorithms and procedures in a novel way.

Keywords: degree of our knowledge, chaotic factor, complex probability set, probability norm, complex random vector, convergence probability, divergence probability, simulation

1. Introduction

> *"Thus, joining the rigor of the demonstrations of science to the uncertainty of fate, and reconciling these two seemingly contradictory things, it can, taking its name from both, appropriately arrogate to itself this astonishing title: the geometry of chance."*
>
> *Blaise Pascal*

> *"You believe in the God who plays dice, and I in complete law and order."*
>
> *Albert Einstein, Letter to Max Born*

> *"Chance is the pseudonym of God when He did not want to sign."*
>
> *Anatole France*

> *"There is a certain Eternal Law, to wit, Reason, existing in the mind of God and governing the whole universe."*
>
> *Saint Thomas Aquinas*

"An equation has no meaning for me unless it expresses a thought of God."
Srinivasa Ramanujan

Calculating probabilities is the crucial task of classical probability theory. Adding supplementary dimensions to nondeterministic experiments will yield a deterministic expression of the theory of probability. This is the novel and original idea at the foundations of my complex probability paradigm. As a matter of fact, probability theory is a stochastic system of axioms in its essence; that means that the phenomena outputs are due to randomness and chance. Adding new imaginary dimensions to the nondeterministic phenomenon happening in the set $\mathcal{R}$ will lead to a deterministic phenomenon, and thus, a probabilistic experiment will have a certain output in the set $\mathcal{C}$ of complex probabilities. If the chaotic experiment becomes fully predictable, then we will be completely capable to foretell the output of random events that occur in the real world in all probabilistic processes. Accordingly, the task that has been achieved here was to extend the set $\mathcal{R}$ of random real probabilities to the deterministic set $\mathcal{C} = \mathcal{R} + \mathcal{M}$ of complex probabilities and this by incorporating the contributions of the set $\mathcal{M}$ which is the set of complementary imaginary probabilities to the set $\mathcal{R}$. Consequently, since this extension reveals to be successful, an innovative paradigm of stochastic sciences and prognostic was put forward in which all nondeterministic phenomena in $\mathcal{R}$ was expressed deterministically in $\mathcal{C}$. I coined this novel model by the term "the complex probability paradigm" that was initiated and established in my 14 earlier research works [1–14].

2. The purpose and the advantages of the current chapter

The advantages and the purpose of the present chapter are to [15–39]:

1. Extend the theory of classical probability to cover the complex numbers set, hence to connect the probability theory to the field of complex analysis and variables. This task was initiated and developed in my earlier 14 works.

2. Apply the novel paradigm and its original probability axioms to Monte Carlo techniques.

3. Prove that all phenomena that are nondeterministic can be transformed to deterministic phenomena in the complex probabilities set which is $\mathcal{C}$.

4. Compute and quantify both the chaotic factor and the degree of our knowledge of Monte Carlo procedures.

5. Represent and show the graphs of the functions and parameters of the innovative model related to Monte Carlo algorithms.

6. Demonstrate that the classical probability concept is permanently equal to 1 in the set of complex probabilities; thus, no chaos, no randomness, no ignorance, no uncertainty, no unpredictability, no nondeterminism, and no disorder exist in

$$\mathcal{C}\text{ (complex set)} = \mathcal{R}\text{ (real set)} + \mathcal{M}\text{ (imaginary set)}.$$

7. Prepare to apply this inventive paradigm to other topics in prognostics and to the field of stochastic processes. These will be the goals of my future research publications.

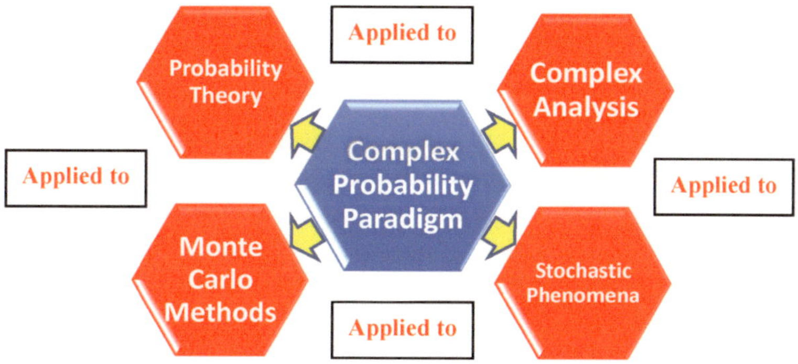

Figure 1.
The diagram of the major aims of the complex probability paradigm.

Regarding some applications of the novel established model and as a subsequent work, it can be applied to any nondeterministic experiments using Monte Carlo algorithms whether in the continuous or in the discrete cases.

Moreover, compared with existing literature, the major contribution of the current chapter is to apply the innovative complex probability paradigm to the techniques and concepts of the probabilistic Monte Carlo simulations and algorithms.

The next figure displays the major aims and purposes of the complex probability paradigm (*CPP*) (**Figure 1**).

3. The complex probability paradigm

3.1 The original Andrey Nikolaevich Kolmogorov system of axioms

The simplicity of Kolmogorov's system of axioms may be surprising [1–14]. Let E be a collection of elements $\{E_1, E_2, \ldots\}$ called elementary events and let F be a set of subsets of E called random events. The five axioms for a finite set E are:

Axiom 1: F is a field of sets.

Axiom 2: F contains the set E.

Axiom 3: A nonnegative real number $P_{rob}(A)$, called the probability of A, is assigned to each set A in F. We have always $0 \le P_{rob}(A) \le 1$.

Axiom 4: $P_{rob}(E)$ equals 1.

Axiom 5: If A and B have no elements in common, the number assigned to their union is

$$P_{rob}(A \cup B) = P_{rob}(A) + P_{rob}(B)$$

hence, we say that A and B are disjoint; otherwise, we have

$$P_{rob}(A \cup B) = P_{rob}(A) + P_{rob}(B) - P_{rob}(A \cap B)$$

And we say also that $P_{rob}(A \cap B) = P_{rob}(A) \times P_{rob}(B/A) = P_{rob}(B) \times P_{rob}(A/B)$ which is the conditional probability. If both A and B are independent then $P_{rob}(A \cap B) = P_{rob}(A) \times P_{rob}(B)$.

Moreover, we can generalize and say that for N disjoint (mutually exclusive) events $A_1, A_2, \ldots, A_j, \ldots, A_N$ (for $1 \le j \le N$), we have the following additivity rule:

$$P_{rob}\left(\bigcup_{j=1}^{N} A_j\right) = \sum_{j=1}^{N} P_{rob}(A_j)$$

And we say also that for N independent events $A_1, A_2, \dots, A_j, \dots, A_N$ (for $1 \leq j \leq N$), we have the following product rule

$$P_{rob}\left(\bigcap_{j=1}^{N} A_j\right) = \prod_{j=1}^{N} P_{rob}(A_j)$$

3.2 Adding the imaginary part $\mathcal{M}$

Now, we can add to this system of axioms an imaginary part such that:

Axiom 6: Let $P_m = i \times (1 - P_r)$ be the probability of an associated complementary event in $\mathcal{M}$ (the imaginary part) to the event A in $\mathcal{R}$ (the real part). It follows that $P_r + P_m/i = 1$ where i is the imaginary number with $i = \sqrt{-1}$ or $i^2 = -1$.

Axiom 7: We construct the complex number or vector $Z = P_r + P_m = P_r + i(1 - P_r)$ having a norm $|Z|$ such that

$$|Z|^2 = P_r^2 + (P_m/i)^2.$$

Axiom 8: Let Pc denotes the probability of an event in the complex probability universe $\mathcal{C}$ where $\mathcal{C} = \mathcal{R} + \mathcal{M}$. We say that Pc is the probability of an event A in $\mathcal{R}$ with its associated event in $\mathcal{M}$ such that

$$Pc^2 = (P_r + P_m/i)^2 = |Z|^2 - 2iP_rP_m \text{ and is always equal to 1.}$$

We can see that by taking into consideration the set of imaginary probabilities, we added three new and original axioms, and consequently the system of axioms

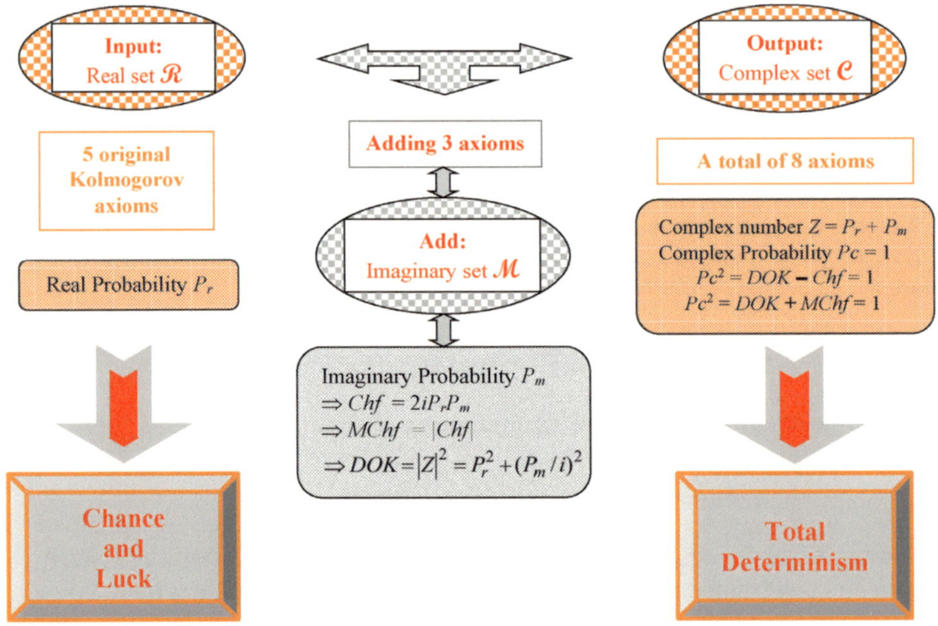

Figure 2.
The EKA *or the* CPP *diagram.*

defined by Kolmogorov was hence expanded to encompass the set of imaginary numbers.

3.3 A brief interpretation of the novel paradigm

To summarize the novel paradigm, we state that in the real probability universe $\mathcal{R}$, our degree of our certain knowledge is undesirably imperfect and hence unsatisfactory; thus, we extend our analysis to the set of complex numbers $\mathcal{C}$ which incorporates the contributions of both the set of real probabilities which is $\mathcal{R}$ and the complementary set of imaginary probabilities which is $\mathcal{M}$. Afterward, this will yield an absolute and perfect degree of our knowledge in the probability universe $\mathcal{C} = \mathcal{R} + \mathcal{M}$ because $Pc = 1$ constantly. As a matter of fact, the work in the universe $\mathcal{C}$ of complex probabilities gives way to a sure forecast of any stochastic experiment, since in $\mathcal{C}$ we remove and subtract from the computed degree of our knowledge the measured chaotic factor. This will generate in the universe $\mathcal{C}$ a probability equal to 1 ($Pc^2 = DOK - Chf = DOK + MChf = 1 = Pc$). Many applications taking into consideration numerous continuous and discrete probability distributions in my 14 previous research papers confirm this hypothesis and innovative paradigm. The extended Kolmogorov axioms (EKA) or the complex probability paradigm (CPP) can be shown and summarized in the next illustration (**Figure 2**).

4. The Monte Carlo techniques and the complex probability paradigm parameters

4.1 The divergence and convergence probabilities

Let R_E be the exact result of the stochastic phenomenon or of a multidimensional or simple integral that are not always possible to compute by probability theory ordinary procedures or by deterministic numerical means or by calculus [1–14]. And let R_A be the phenomenon and integrals approximate results calculated by the techniques of Monte Carlo:

The relative error in the Monte Carlo methods is $\text{Rel.Error} = \left|\frac{R_E - R_A}{R_E}\right| = \left|1 - \frac{R_A}{R_E}\right|$.

Additionally, the percent relative error is = $100\% \times \left|\frac{R_E - R_A}{R_E}\right|$ and is always between 0% and 100%. Therefore, the relative error is always between 0 and 1. Hence

$$0 \le \left|\frac{R_E - R_A}{R_E}\right| \le 1 \Leftrightarrow \begin{cases} 0 \le \left(\dfrac{R_E - R_A}{R_E}\right) \le 1 & \text{if } R_A \le R_E \\ 0 \le -\left(\dfrac{R_E - R_A}{R_E}\right) \le 1 & \text{if } R_A \ge R_E \end{cases} \Leftrightarrow \begin{cases} 0 \le R_A \le R_E \\ R_E \le R_A \le 2R_E \end{cases}$$

Moreover, we define the real probability in the set $\mathcal{R}$ by

$$P_r = 1 - \left|\frac{R_E - R_A}{R_E}\right| = 1 - \left|1 - \frac{R_A}{R_E}\right| = \begin{cases} 1 - \left(1 - \dfrac{R_A}{R_E}\right) & \text{if } 0 \le R_A \le R_E \\ 1 + \left(1 - \dfrac{R_A}{R_E}\right) & \text{if } R_E \le R_A \le 2R_E \end{cases}$$

$$= \begin{cases} \dfrac{R_A}{R_E} & \text{if } 0 \le R_A \le R_E \\ 2 - \dfrac{R_A}{R_E} & \text{if } R_E \le R_A \le 2R_E \end{cases}$$

= 1 – the relative error in the Monte Carlo method.
= probability of Monte Carlo method convergence in $\mathcal{R}$.
And therefore,

$$P_m = i(1-P_r) = i\left\{1-\left[1-\left|\frac{R_E-R_A}{R_E}\right|\right]\right\} = i\left\{1-\left[1-\left|1-\frac{R_A}{R_E}\right|\right]\right\} = i\left|1-\frac{R_A}{R_E}\right|$$

$$= \begin{cases} i\left(1-\frac{R_A}{R_E}\right) & \text{if } 0 \le R_A \le R_E \\ -i\left(1-\frac{R_A}{R_E}\right) & \text{if } R_E \le R_A \le 2R_E \end{cases} = \begin{cases} i\left(1-\frac{R_A}{R_E}\right) & \text{if } 0 \le R_A \le R_E \\ i\left(\frac{R_A}{R_E}-1\right) & \text{if } R_E \le R_A \le 2R_E \end{cases}$$

= probability of Monte Carlo method divergence in the imaginary complementary probability set $\mathcal{M}$ since it is the imaginary complement of P_r.
Consequently,

$$P_m/i = 1-P_r = \left|1-\frac{R_A}{R_E}\right| = \begin{cases} 1-\frac{R_A}{R_E} & \text{if } 0 \le R_A \le R_E \\ \frac{R_A}{R_E}-1 & \text{if } R_E \le R_A \le 2R_E \end{cases}$$

= the relative error in the Monte Carlo method.
= probability of Monte Carlo method divergence in $\mathcal{R}$ since it is the real complement of P_r.

In the case where $0 \le R_A \le R_E \Rightarrow 0 \le \frac{R_A}{R_E} \le 1 \Rightarrow 0 \le P_r \le 1$ and we deduce also that $0 \le \left(1-\frac{R_A}{R_E}\right) \le 1 \Rightarrow 0 \le P_m/i \le 1$ and $\Rightarrow 0 \le P_m \le i$.

And in the case where $R_E \le R_A \le 2R_E \Rightarrow 1 \le \frac{R_A}{R_E} \le 2 \Rightarrow 0 \le \left(2-\frac{R_A}{R_E}\right) \le 1 \Rightarrow$ $0 \le P_r \le 1$ and we deduce also that $0 \le \left(\frac{R_A}{R_E}-1\right) \le 1 \Rightarrow 0 \le P_m/i \le 1$ and $\Rightarrow 0 \le P_m \le i$.

Consequently, if $R_A = 0$ or $R_A = 2R_E$ that means before the beginning of the simulation, then

$$P_r = P_{rob}\text{ (convergence) in } \mathcal{R} = 0$$
$$P_m = P_{rob}\text{ (divergence) in } \mathcal{M} = i$$
$$P_m/i = P_{rob}\text{ (divergence) in } \mathcal{R} = 1$$

And if $R_A = R_E$ that means at the end of Monte Carlo simulation, then

$$P_r = P_{rob}\text{ (convergence) in } \mathcal{R} = 1$$
$$P_m = P_{rob}\text{ (divergence) in } \mathcal{M} = 0$$
$$P_m/i = P_{rob}\text{ (divergence) in } \mathcal{R} = 0$$

4.2 The complex random vector Z in $\mathcal{C} = \mathcal{R} + \mathcal{M}$

We have
$$Z = P_r + P_m = \begin{cases} \frac{R_A}{R_E} + i\left(1-\frac{R_A}{R_E}\right) & \text{if } 0 \le R_A \le R_E \\ \left(2-\frac{R_A}{R_E}\right) + i\left(\frac{R_A}{R_E}-1\right) & \text{if } R_E \le R_A \le 2R_E \end{cases}.$$
$$= \text{Re}\,(Z) + i\text{Im}(Z)$$

where

$$\mathrm{Re}\,(Z) = P_r = \begin{cases} \dfrac{R_A}{R_E} & \text{if } 0 \le R_A \le R_E \\ 2 - \dfrac{R_A}{R_E} & \text{if } R_E \le R_A \le 2R_E \end{cases} = \text{the real part of } Z$$

and

$$\mathrm{Im}(Z) = P_m/i = \begin{cases} 1 - \dfrac{R_A}{R_E} & \text{if } 0 \le R_A \le R_E \\ \dfrac{R_A}{R_E} - 1 & \text{if } R_E \le R_A \le 2R_E \end{cases} = \text{the imaginary part of } Z.$$

That means that the complex random vector Z is the sum in $\mathcal{C}$ of the real probability of convergence in $\mathcal{R}$ and of the imaginary probability of divergence in $\mathcal{M}$.

If $R_A = 0$ or $R_A = 2R_E$ (before the simulation begins), then

$$P_r = \frac{R_A}{R_E} = \frac{0}{R_E} = 0 \text{ or } P_r = 2 - \frac{R_A}{R_E} = 2 - \frac{2R_E}{R_E} = 2 - 2 = 0.$$

and

$$P_m = i\left(1 - \frac{R_A}{R_E}\right) = i\left(1 - \frac{0}{R_E}\right) = i(1-0) = i \text{ or } P_m = i\left(\frac{R_A}{R_E} - 1\right) = i\left(\frac{2R_E}{R_E} - 1\right)$$
$$= i(2-1) = i$$

therefore $Z = 0 + i = i$.

If $R_A = \frac{R_E}{2}$ or $R_A = \frac{3R_E}{2}$ (at the middle of the simulation), then

$$P_r = \begin{cases} \dfrac{R_A}{R_E} & \text{if } 0 \le R_A \le R_E \\ 2 - \dfrac{R_A}{R_E} & \text{if } R_E \le R_A \le 2R_E \end{cases} = \begin{cases} \dfrac{R_E}{2R_E} = 0.5 & \text{if } 0 \le R_A \le R_E \\ 2 - \dfrac{3R_E}{2R_E} = 0.5 & \text{if } R_E \le R_A \le 2R_E \end{cases} \Leftrightarrow P_r = 0.5$$

and

$$P_m = \begin{cases} i\left(1 - \dfrac{R_A}{R_E}\right) & \text{if } 0 \le R_A \le R_E \\ i\left(\dfrac{R_A}{R_E} - 1\right) & \text{if } R_E \le R_A \le 2R_E \end{cases}$$
$$= \begin{cases} i\left(1 - \dfrac{R_E}{2R_E}\right) = 0.5i & \text{if } 0 \le R_A \le R_E \\ i\left(\dfrac{3R_E}{2R_E} - 1\right) = 0.5i & \text{if } R_E \le R_A \le 2R_E \end{cases} \Leftrightarrow P_m = 0.5i$$

therefore $Z = 0.5 + 0.5i$.

If $R_A = R_E$ (at the simulation end), then

$$P_r = \begin{cases} \dfrac{R_A}{R_E} = \dfrac{R_E}{R_E} = 1 & \text{if } 0 \le R_A \le R_E \\ 2 - \dfrac{R_A}{R_E} = 2 - \dfrac{R_E}{R_E} = 2 - 1 = 1 & \text{if } R_E \le R_A \le 2R_E \end{cases} \Leftrightarrow P_r = 1$$

and

$$P_m = \begin{cases} i\left(1-\dfrac{R_A}{R_E}\right) & \text{if } 0 \le R_A \le R_E \\ i\left(\dfrac{R_A}{R_E}-1\right) & \text{if } R_E \le R_A \le 2R_E \end{cases} = \begin{cases} i\left(1-\dfrac{R_E}{R_E}\right) & \text{if } 0 \le R_A \le R_E \\ i\left(\dfrac{R_E}{R_E}-1\right) & \text{if } R_E \le R_A \le 2R_E \end{cases}$$

$$= \begin{cases} 0 & \text{if } 0 \le R_A \le R_E \\ 0 & \text{if } R_E \le R_A \le 2R_E \end{cases}$$

$$\Leftrightarrow P_m = 0$$

therefore $Z = 1 + 0i = 1$.

4.3 The degree of our knowledge, *DOK*

We have

$$DOK = |Z|^2 = P_r^2 + (P_m/i)^2 = \begin{cases} \left(\frac{R_A}{R_E}\right)^2 & \text{if } 0 \le R_A \le R_E \\ \left(2-\frac{R_A}{R_E}\right)^2 & \text{if } R_E \le R_A \le 2R_E \end{cases} + \begin{cases} \left(1-\frac{R_A}{R_E}\right)^2 & \text{if } 0 \le R_A \le R_E \\ \left(\frac{R_A}{R_E}-1\right)^2 & \text{if } R_E \le R_A \le 2R_E \end{cases}$$

$$= \begin{cases} \left(\frac{R_A}{R_E}\right)^2 + \left(1-\frac{R_A}{R_E}\right)^2 & \text{if } 0 \le R_A \le R_E \\ \left(2-\frac{R_A}{R_E}\right)^2 + \left(\frac{R_A}{R_E}-1\right)^2 & \text{if } R_E \le R_A \le 2R_E \end{cases} = \begin{cases} 2\left(\frac{R_A}{R_E}\right)^2 - 2\left(\dfrac{R_A}{R_E}\right) + 1 & \text{if } 0 \le R_A \le R_E \\ 2\left(\frac{R_A}{R_E}\right)^2 - 6\left(\dfrac{R_A}{R_E}\right) + 5 & \text{if } R_E \le R_A \le 2R_E \end{cases}$$

From *CPP* we have that $0.5 \le DOK \le 1$, then if $DOK = 0.5$

$$\Leftrightarrow \begin{cases} 2\left(\frac{R_A}{R_E}\right)^2 - 2\left(\dfrac{R_A}{R_E}\right) + 1 = 0.5 & \text{if } 0 \le R_A \le R_E \\ 2\left(\frac{R_A}{R_E}\right)^2 - 6\left(\dfrac{R_A}{R_E}\right) + 5 = 0.5 & \text{if } R_E \le R_A \le 2R_E \end{cases}$$

then solving the second-degree equations for $\frac{R_A}{R_E}$ gives

$$\begin{cases} \dfrac{R_A}{R_E} = 1/2 & \text{if } 0 \le R_A \le R_E \\ \dfrac{R_A}{R_E} = 3/2 & \text{if } R_E \le R_A \le 2R_E \end{cases} \Leftrightarrow \begin{cases} R_A = R_E/2 & \text{if } 0 \le R_A \le R_E \\ R_A = 3R_E/2 & \text{if } R_E \le R_A \le 2R_E \end{cases} \text{ and vice versa.}$$

That means that *DOK* is minimum when the approximate result R_A is equal to half of the exact result R_E if $0 \le R_A \le R_E$ or when the approximate result is equal to three times the half of the exact result if $R_E \le R_A \le 2R_E$, which means at the middle of the simulation.

In addition, if $DOK = 1$, then

$$\Leftrightarrow \begin{cases} 2\left(\frac{R_A}{R_E}\right)^2 - 2\left(\dfrac{R_A}{R_E}\right) + 1 = 1 & \text{if } 0 \le R_A \le R_E \\ 2\left(\frac{R_A}{R_E}\right)^2 - 6\left(\dfrac{R_A}{R_E}\right) + 5 = 1 & \text{if } R_E \le R_A \le 2R_E \end{cases} \Leftrightarrow \begin{cases} \left(\frac{R_A}{R_E}\right)^2 - \left(\dfrac{R_A}{R_E}\right) = 0 & \text{if } 0 \le R_A \le R_E \\ 2\left(\frac{R_A}{R_E}\right)^2 - 6\left(\dfrac{R_A}{R_E}\right) + 4 = 0 & \text{if } R_E \le R_A \le 2R_E \end{cases}$$

$$\Leftrightarrow \begin{cases} R_A = 0 \text{ OR } R_A = R_E & \text{if } 0 \le R_A \le R_E \\ R_A = 2R_E \text{ OR } R_A = R_E & \text{if } R_E \le R_A \le 2R_E \end{cases} \text{ and vice versa.}$$

That means that *DOK* is maximum when the approximate result R_A is equal to 0 or $2R_E$ (before the beginning of the simulation) and when it is equal to the exact result R_E (at the end of the simulation). We can deduce that we have perfect and total knowledge of the stochastic experiment before the beginning of Monte Carlo simulation since no randomness was introduced yet, as well as at the end of the simulation after the convergence of the method to the exact result.

4.4 The chaotic factor, *Chf*

We have

$$\begin{aligned} Chf &= 2iP_rP_m \\ &= 2i \times \begin{cases} \dfrac{R_A}{R_E} & \text{if } 0 \le R_A \le R_E \\ 2 - \dfrac{R_A}{R_E} & \text{if } R_E \le R_A \le 2R_E \end{cases} \times \begin{cases} i\left(1 - \dfrac{R_A}{R_E}\right) & \text{if } 0 \le R_A \le R_E \\ i\left(\dfrac{R_A}{R_E} - 1\right) & \text{if } R_E \le R_A \le 2R_E \end{cases} \end{aligned}$$

Since $i^2 = -1$ then

$$Chf = \begin{cases} -2\left(\dfrac{R_A}{R_E}\right)\left(1 - \dfrac{R_A}{R_E}\right) & \text{if } 0 \le R_A \le R_E \\ -2\left(2 - \dfrac{R_A}{R_E}\right)\left(\dfrac{R_A}{R_E} - 1\right) & \text{if } R_E \le R_A \le 2R_E \end{cases}$$

From *CPP* we have that $-0.5 \le Chf \le 0$, and then if $Chf = -0.5$

$$\Leftrightarrow \begin{cases} -2\left(\dfrac{R_A}{R_E}\right)\left(1 - \dfrac{R_A}{R_E}\right) = -0.5 & \text{if } 0 \le R_A \le R_E \\ -2\left(2 - \dfrac{R_A}{R_E}\right)\left(\dfrac{R_A}{R_E} - 1\right) = -0.5 & \text{if } R_E \le R_A \le 2R_E \end{cases} \Leftrightarrow \begin{cases} R_A = R_E/2 & \text{if } 0 \le R_A \le R_E \\ R_A = 3R_E/2 & \text{if } R_E \le R_A \le 2R_E \end{cases}$$

and vice versa.

That means that *Chf* is minimum when the approximate result R_A is equal to half of the exact result R_E if $0 \le R_A \le R_E$ or when the approximate result is equal to three times the half of the exact result if $R_E \le R_A \le 2R_E$, which means at the middle of the simulation.

In addition, if $Chf = 0$ then

$$\Leftrightarrow \begin{cases} -2\left(\dfrac{R_A}{R_E}\right)\left(1 - \dfrac{R_A}{R_E}\right) = 0 & \text{if } 0 \le R_A \le R_E \\ -2\left(2 - \dfrac{R_A}{R_E}\right)\left(\dfrac{R_A}{R_E} - 1\right) = 0 & \text{if } R_E \le R_A \le 2R_E \end{cases} \Leftrightarrow \begin{cases} R_A = 0 \text{ OR } R_A = R_E & \text{if } 0 \le R_A \le R_E \\ R_A = 2R_E \text{ OR } R_A = R_E & \text{if } R_E \le R_A \le 2R_E \end{cases}$$

And, conversely, if $\begin{cases} R_A = 0 \text{ OR } R_A = R_E & \text{if } 0 \le R_A \le R_E \\ R_A = 2R_E \text{ OR } R_A = R_E & \text{if } R_E \le R_A \le 2R_E \end{cases}$, then $Chf = 0$.

That means that *Chf* is equal to 0 when the approximate result R_A is equal to 0 or $2R_E$ (before the beginning of the simulation) and when it is equal to the exact result R_E (at the end of the simulation).

4.5 The magnitude of the chaotic factor, *MChf*

We have

$$MChf = |Chf| = -2iP_rP_m = -2i \times \begin{cases} \dfrac{R_A}{R_E} & \text{if } 0 \le R_A \le R_E \\ 2 - \dfrac{R_A}{R_E} & \text{if } R_E \le R_A \le 2R_E \end{cases} \times \begin{cases} i\left(1 - \dfrac{R_A}{R_E}\right) & \text{if } 0 \le R_A \le R_E \\ i\left(\dfrac{R_A}{R_E} - 1\right) & \text{if } R_E \le R_A \le 2R_E \end{cases}$$

Since $i^2 = -1$ then

$$MChf = \begin{cases} 2\left(\dfrac{R_A}{R_E}\right)\left(1 - \dfrac{R_A}{R_E}\right) & \text{if } 0 \le R_A \le R_E \\ 2\left(2 - \dfrac{R_A}{R_E}\right)\left(\dfrac{R_A}{R_E} - 1\right) & \text{if } R_E \le R_A \le 2R_E \end{cases}$$

From *CPP* we have that $0 \le MChf \le 0.5$, and then if $MChf = 0.5$

$$\Leftrightarrow \begin{cases} 2\left(\dfrac{R_A}{R_E}\right)\left(1 - \dfrac{R_A}{R_E}\right) = 0.5 & \text{if } 0 \le R_A \le R_E \\ 2\left(2 - \dfrac{R_A}{R_E}\right)\left(\dfrac{R_A}{R_E} - 1\right) = 0.5 & \text{if } R_E \le R_A \le 2R_E \end{cases} \Leftrightarrow \begin{cases} R_A = R_E/2 & \text{if } 0 \le R_A \le R_E \\ R_A = 3R_E/2 & \text{if } R_E \le R_A \le 2R_E \end{cases}$$

and vice versa.

That means that *MChf* is maximum when the approximate result R_A is equal to half of the exact result R_E if $0 \le R_A \le R_E$ or when the approximate result is equal to three times the half of the exact result if $R_E \le R_A \le 2R_E$, which means at the middle of the simulation. This implies that the magnitude of the chaos (*MChf*) introduced by the random variables used in Monte Carlo method is maximum at the halfway of the simulation.

In addition, if $MChf = 0$, then

$$\Leftrightarrow \begin{cases} 2\left(\dfrac{R_A}{R_E}\right)\left(1 - \dfrac{R_A}{R_E}\right) = 0 & \text{if } 0 \le R_A \le R_E \\ 2\left(2 - \dfrac{R_A}{R_E}\right)\left(\dfrac{R_A}{R_E} - 1\right) = 0 & \text{if } R_E \le R_A \le 2R_E \end{cases} \Leftrightarrow \begin{cases} R_A = 0 \text{ OR } R_A = R_E & \text{if } 0 \le R_A \le R_E \\ R_A = 2R_E \text{ OR } R_A = R_E & \text{if } R_E \le R_A \le 2R_E \end{cases}$$

And, conversely, if $\begin{cases} R_A = 0 \text{ OR } R_A = R_E & \text{if } 0 \le R_A \le R_E \\ R_A = 2R_E \text{ OR } R_A = R_E & \text{if } R_E \le R_A \le 2R_E \end{cases}$, then $MChf = 0$.

That means that *MChf* is minimum and is equal to 0 when the approximate result R_A is equal to 0 or $2R_E$ (before the beginning of the simulation) and when it is equal to the exact result R_E (at the end of the simulation). We can deduce that the magnitude of the chaos in the stochastic experiment is null before the beginning of Monte Carlo simulation since no randomness was introduced yet, as well as at the end of the simulation after the convergence of the method to the exact result when randomness has finished its task in the stochastic Monte Carlo method and experiment.

4.6 The probability Pc in the probability set $\mathcal{C} = \mathcal{R} + \mathcal{M}$

We have

$$Pc^2 = DOK - Chf = DOK + MChf$$

$$= \begin{cases} 2\left(\frac{R_A}{R_E}\right)^2 - 2\left(\frac{R_A}{R_E}\right) + 1 & \text{if } 0 \le R_A \le R_E \\ 2\left(\frac{R_A}{R_E}\right)^2 - 6\left(\frac{R_A}{R_E}\right) + 5 & \text{if } R_E \le R_A \le 2R_E \end{cases} - \begin{cases} -2\left(\frac{R_A}{R_E}\right)\left(1 - \frac{R_A}{R_E}\right) & \text{if } 0 \le R_A \le R_E \\ -2\left(2 - \frac{R_A}{R_E}\right)\left(\frac{R_A}{R_E} - 1\right) & \text{if } R_E \le R_A \le 2R_E \end{cases}$$

$$= \begin{cases} 1 & \text{if } 0 \le R_A \le R_E \\ 1 & \text{if } R_E \le R_A \le 2R_E \end{cases} \Leftrightarrow Pc^2 = 1 \quad \text{for } 0 \le \forall R_A \le 2R_E$$

$\Leftrightarrow Pc = 1$ = probability of convergence in $\mathcal{C}$; therefore,

$$Pc = \begin{cases} \frac{R_A}{R_E} = 1 & \text{if } 0 \le R_A \le R_E \\ 2 - \frac{R_A}{R_E} = 1 & \text{if } R_E \le R_A \le 2R_E \end{cases} \Leftrightarrow \begin{cases} R_A = R_E & \text{if } 0 \le R_A \le R_E \\ R_A = R_E & \text{if } R_E \le R_A \le 2R_E \end{cases}$$

$\Leftrightarrow R_A = R_E \quad \text{for } 0 \le \forall R_A \le 2R_E$ continuously in the probability set $\mathcal{C} = \mathcal{R} + \mathcal{M}$. This is due to the fact that in $\mathcal{C}$, we have subtracted in the equation above the chaotic factor *Chf* from our knowledge *DOK*, and therefore we have eliminated chaos caused and introduced by all the random variables and the stochastic fluctuations that lead to approximate results in the Monte Carlo simulation in $\mathcal{R}$. Therefore, since in $\mathcal{C}$ we have always $R_A = R_E$, then the Monte Carlo simulation which is a stochastic method by nature in $\mathcal{R}$ becomes after applying the *CPP* a deterministic method in $\mathcal{C}$ since the probability of convergence of any random experiment in $\mathcal{C}$ is constantly and permanently equal to 1 for any iterations number *N*.

4.7 The rates of change of the probabilities in $\mathcal{R}$, $\mathcal{M}$, and $\mathcal{C}$

Since

$$Z = P_r + P_m = \begin{cases} \frac{R_A}{R_E} + i\left(1 - \frac{R_A}{R_E}\right) & \text{if } 0 \le R_A \le R_E \\ \left(2 - \frac{R_A}{R_E}\right) + i\left(\frac{R_A}{R_E} - 1\right) & \text{if } R_E \le R_A \le 2R_E \end{cases} = \mathrm{Re}(Z) + i\mathrm{Im}(Z)$$

Then

$$\frac{dZ}{dR_A} = \frac{dP_r}{dR_A} + \frac{dP_m}{dR_A} = \begin{cases} \frac{d}{dR_A}\left[\frac{R_A}{R_E} + i\left(1 - \frac{R_A}{R_E}\right)\right] & \text{if } 0 \le R_A \le R_E \\ \frac{d}{dR_A}\left[\left(2 - \frac{R_A}{R_E}\right) + i\left(\frac{R_A}{R_E} - 1\right)\right] & \text{if } R_E \le R_A \le 2R_E \end{cases}$$

$$= \begin{cases} \frac{d}{dR_A}\left[\frac{R_A}{R_E}\right] + \frac{d}{dR_A}\left[i\left(1 - \frac{R_A}{R_E}\right)\right] & \text{if } 0 \le R_A \le R_E \\ \frac{d}{dR_A}\left[2 - \frac{R_A}{R_E}\right] + \frac{d}{dR_A}\left[i\left(\frac{R_A}{R_E} - 1\right)\right] & \text{if } R_E \le R_A \le 2R_E \end{cases}$$

$$= \begin{cases} \frac{1}{R_E} - \frac{i}{R_E} = \frac{1}{R_E}(1 - i) & \text{if } 0 \le R_A \le R_E \\ -\frac{1}{R_E} + \frac{i}{R_E} = \frac{1}{R_E}(i - 1) & \text{if } R_E \le R_A \le 2R_E \end{cases}$$

Therefore,

$$\text{Re}\left[\frac{dZ}{dR_A}\right] = \frac{dP_r}{dR_A} = \begin{cases} +\frac{1}{R_E} & \text{if } 0 \le R_A \le R_E \\ -\frac{1}{R_E} & \text{if } R_E \le R_A \le 2R_E \end{cases}$$

$$= \begin{cases} \text{constant} > 0 & \text{if } 0 \le R_A \le R_E \text{ and } R_E > 0 \\ \text{constant} < 0 & \text{if } R_E \le R_A \le 2R_E \text{ and } R_E > 0 \end{cases}$$

That means that the slope of the probability of convergence in $\mathcal{R}$ or its rate of change is constant and positive if $0 \le R_A \le R_E$, and constant and negative if $R_E \le R_A \le 2R_E$, and it depends only on R_E; hence, we have a constant increase in P_r (the convergence probability) as a function of the iterations number N as R_A increases from 0 to R_E and as R_A decreases from $2R_E$ to R_E till P_r reaches the value 1 that means till the random experiment converges to R_E:

$$\text{Im}\left[\frac{dZ}{dR_A}\right] = \frac{1}{i}\frac{dP_m}{dR_A} = \frac{d(P_m/i)}{dR_A} = \begin{cases} -\frac{1}{R_E} & \text{if } 0 \le R_A \le R_E \\ +\frac{1}{R_E} & \text{if } R_E \le R_A \le 2R_E \end{cases}$$

$$= \begin{cases} \text{constant} < 0 & \text{if } 0 \le R_A \le R_E \text{ and } R_E > 0 \\ \text{constant} > 0 & \text{if } R_E \le R_A \le 2R_E \text{ and } R_E > 0 \end{cases}$$

That means that the slopes of the probabilities of divergence in $\mathcal{R}$ and $\mathcal{M}$ or their rates of change are constant and negative if $0 \le R_A \le R_E$ and constant and positive if $R_E \le R_A \le 2R_E$ and they depend only on R_E; hence, we have a constant decrease in P_m/i and P_m (the divergence probabilities) as functions of the iterations number N as R_A increases from 0 to R_E and as R_A decreases from $2R_E$ to R_E till P_m/i and P_m reach the value 0 that means till the random experiment converges to R_E.

Additionally,

$$\left|\frac{dZ}{dR_A}\right|^2 = \left[\frac{dP_r}{dR_A}\right]^2 + \left[\frac{1}{i}\frac{dP_m}{dR_A}\right]^2 = \left[\frac{dP_r}{dR_A}\right]^2 + \left[\frac{d(P_m/i)}{dR_A}\right]^2$$

$$= \begin{cases} \left(\frac{1}{R_E}\right)^2 + \left(-\frac{1}{R_E}\right)^2 & \text{if } 0 \le R_A \le R_E \\ \left(-\frac{1}{R_E}\right)^2 + \left(\frac{1}{R_E}\right)^2 & \text{if } R_E \le R_A \le 2R_E \end{cases}$$

$$\Leftrightarrow \left|\frac{dZ}{dR_A}\right|^2 = \frac{1}{R_E^2} + \frac{1}{R_E^2} = \frac{2}{R_E^2} \quad \text{for } 0 \le \forall R_A \le 2R_E$$

$$\Leftrightarrow \left|\frac{dZ}{dR_A}\right| = \frac{\sqrt{2}}{R_E} = \text{constant} > 0 \ \text{ if } \ R_E > 0;$$

that means that the module of the slope of the complex probability vector Z in $\mathcal{C}$ or of its rate of change is constant and positive and it depends only on R_E; hence, we have a constant increase in $\text{Re}\,(Z)$ and a constant decrease in $\text{Im}(Z)$ as functions of the iterations number N and as Z goes from $(0, i)$ at $N = 0$ till $(1,0)$ at the simulation end; hence, till $\text{Re}\,(Z) = P_r$ reaches the value 1 that means till the random experiment converges to R_E.

Furthermore, since $Pc^2 = DOK - Chf = DOK + MChf = 1$, then $Pc = 1$ = probability of convergence in $\mathcal{C}$, and consequently

$$\frac{d(Pc)}{dR_A} = \frac{d(1)}{dR_A} = 0,$$

which means that Pc is constantly equal to 1 for every value of R_A, of R_E, and of the iterations number N, which means for any stochastic experiment and for any simulation of Monte Carlo method. So, we conclude that in $\mathcal{C} = \mathcal{R} + \mathcal{M}$, we have complete and perfect knowledge of the random experiment which has become now a deterministic one since the extension in the complex probability plane $\mathcal{C}$ defined by the *CPP* axioms has changed all stochastic variables to deterministic variables.

5. The new paradigm parameter evaluation

We can infer from what has been developed earlier the following:

The real probability of convergence $P_r(N) = 1 - \left|\frac{R_E - R_A(N)}{R_E}\right|$.

We have $0 \le N \le N_C$ where $N = 0$ corresponds to the instant before the beginning of the random experiment when $R_A(N = 0) = 0$ or $= 2R_E$ and where $N = N_C$ (iterations number needed for the method convergence) corresponds to the instant at the end of the random experiments and Monte Carlo methods when $R_A(N = N_C) \to R_E$.

The imaginary complementary probability of divergence $P_m(N) = i\left|\frac{R_E - R_A(N)}{R_E}\right|$.

The real complementary probability of divergence $P_m(N)/i = \left|\frac{R_E - R_A(N)}{R_E}\right|$.

The random vector of complex probability

$$Z(N) = P_r(N) + P_m(N) = \left[1 - \left|\frac{R_E - R_A(N)}{R_E}\right|\right] + i\left|\frac{R_E - R_A(N)}{R_E}\right|$$

The degree of our knowledge

$$DOK(N) = |Z(N)|^2 = P_r^2(N) + [P_m(N)/i]^2 = \left[1 - \left|\frac{R_E - R_A(N)}{R_E}\right|\right]^2 + \left[\left|\frac{R_E - R_A(N)}{R_E}\right|\right]^2$$

$$= 1 + 2iP_r(N)P_m(N) = 1 - 2P_r(N)[1 - P_r(N)] = 1 - 2P_r(N) + 2P_r^2(N)$$

$$= 1 - 2\left|\frac{R_E - R_A(N)}{R_E}\right| + 2\left[\frac{R_E - R_A(N)}{R_E}\right]^2.$$

$DOK(N)$ is equal to 1 when $P_r(N) = P_r(0) = 0$ and when $P_r(N) = P_r(N_C) = 1$.
The Chaotic factor

$$Chf(N) = 2iP_r(N)P_m(N) = -2P_r(N)[1 - P_r(N)] = -2P_r(N) + 2P_r^2(N)$$

$$= -2\left|\frac{R_E - R_A(N)}{R_E}\right| + 2\left[\frac{R_E - R_A(N)}{R_E}\right]^2$$

$Chf(N)$ is null when $P_r(N) = P_r(0) = 0$ and when $P_r(N) = P_r(N_C) = 1$.
The magnitude of the chaotic factor $MChf$

$$MChf(N) = |Chf(N)| = \quad 2iP_r(N)P_m(N) = 2P_r(N)[1 - P_r(N)] = 2P_r(N) - 2P_r^2(N)$$

$$= 2\left|\frac{R_E - R_A(N)}{R_E}\right| - 2\left[\frac{R_E - R_A(N)}{R_E}\right]^2$$

$MChf(N)$ is null when $P_r(N) = P_r(0) = 0$ and when $P_r(N) = P_r(N_C) = 1$.
At any iteration number N $0 \le \forall N \le N_C$, the probability calculated in the set $\mathcal{C}$ of complex probabilities is as follows:

$$Pc^2(N) = [P_r(N) + P_m(N)/i]^2 = |Z(N)|^2 - 2iP_r(N)P_m(N) = DOK(N) - Chf(N)$$

$$= DOK(N) + MChf(N) = 1$$

then

$$Pc^2(N) = [P_r(N) + P_m(N)/i]^2 = \{P_r(N) + [1 - P_r(N)]\}^2 = 1^2 = 1 \Leftrightarrow Pc(N) = 1 (\text{continuously}).$$

Thus, the prediction in the set $\mathcal{C}$ of the probabilities of convergence of the random Monte Carlo methods is always certain.

Let us consider afterward a multidimensional integral and a stochastic experiment to simulate the Monte Carlo procedures and to quantify, to draw, as well as to visualize all the prognostic and *CPP* parameters.

6. The flowchart of the prognostic model of Monte Carlo techniques and *CPP*

The flowchart that follows illustrates all the procedures of the elaborated prognostic model of *CPP*.

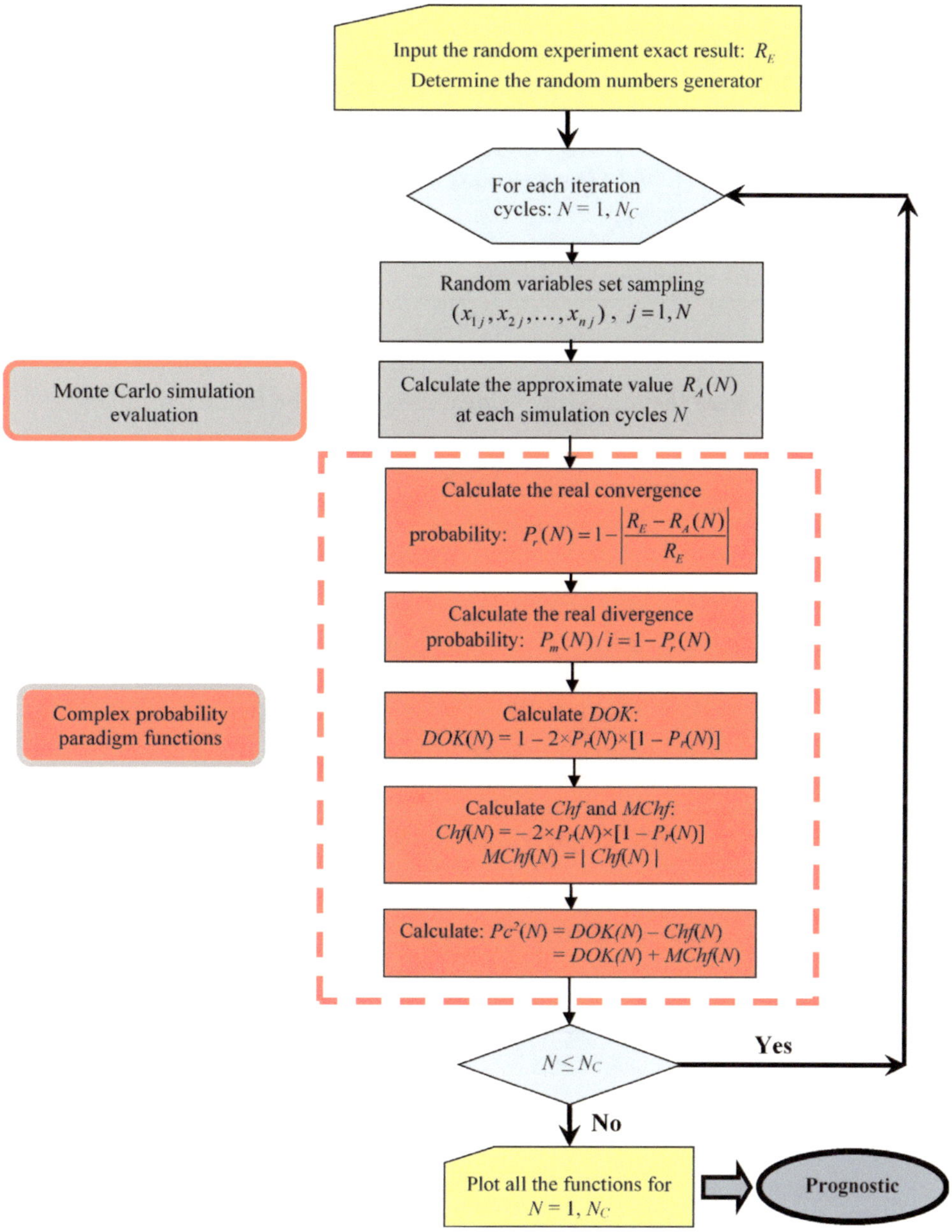

7. Simulation of the new paradigm

Note that all the numerical values found in the simulations of the new paradigm for any iteration cycles N were computed using the 64-bit MATLAB version 2020 software and compared to the values found by Microsoft Visual C++ programs. Additionally, the reader should be careful of the truncation and rounding errors since we represent all numerical values by at most five significant digits and since we are using Monte Carlo techniques of simulation and integration which yield approximate results under the influence of stochastic aspects and variations. We have considered for this purpose a high-capacity computer system: a workstation computer with parallel microprocessors, a 64-bit operating system, and a 64-GB RAM.

7.1 The continuous random case: a four-dimensional multiple integral

The Monte Carlo technique of integration can be summarized by the following equation:

$$\int_{a_1}^{b_1}\int_{a_2}^{b_2}\ldots\int_{a_n}^{b_n} f(x_1,x_2,\ldots,x_n).dx_1dx_2\ldots dx_n \cong$$

$$\frac{[(b_1-a_1)\times(b_2-a_2)\times\ldots\times(b_n-a_n)]}{N}\sum_{j=1}^{N} f(x_{1j},x_{2j},\ldots,x_{nj})$$

Let us consider here the multidimensional integral of the following function:

$$\int_0^{4/3}\int_0^{4/3}\int_0^{4/3}\int_0^{4/3} xyzw.dxdydzdw = \int_0^{4/3}\int_0^{4/3}\int_0^{4/3}\left[\frac{x^2}{2}\right]_0^{4/3} yzw.dydzdw = \int_0^{4/3}\int_0^{4/3}\int_0^{4/3}\frac{16}{18}yzw.dydzdw$$

$$= \frac{8}{9}\int_0^{4/3}\int_0^{4/3}\left[\frac{y^2}{2}\right]_0^{4/3} zw.dzdw = \frac{8}{9}\int_0^{4/3}\int_0^{4/3}\frac{16}{18}zw.dzdw = \frac{64}{81}\int_0^{4/3}\left[\frac{z^2}{2}\right]_0^{4/3} w.dw = \frac{64}{81}\int_0^{4/3}\frac{16}{18}w.dw$$

$$= \frac{512}{729}\left[\frac{w^2}{2}\right]_0^{4/3} = \frac{512}{729}\times\frac{16}{18} = \frac{512}{729}\times\frac{8}{9} = \frac{4,096}{6,561} = 0.62429507696997411\ldots$$

$\Leftrightarrow R_E = 0.62429507696997411\ldots$ by the deterministic methods of calculus.
$\Leftrightarrow f(x,y,z,w) = xyzw$, where x, y, z, and w follow a discrete uniform distribution U such that

$$x \mapsto U(0,4/3), y \mapsto U(0,4/3), z \mapsto U(0,4/3), w \mapsto U(0,4/3)$$

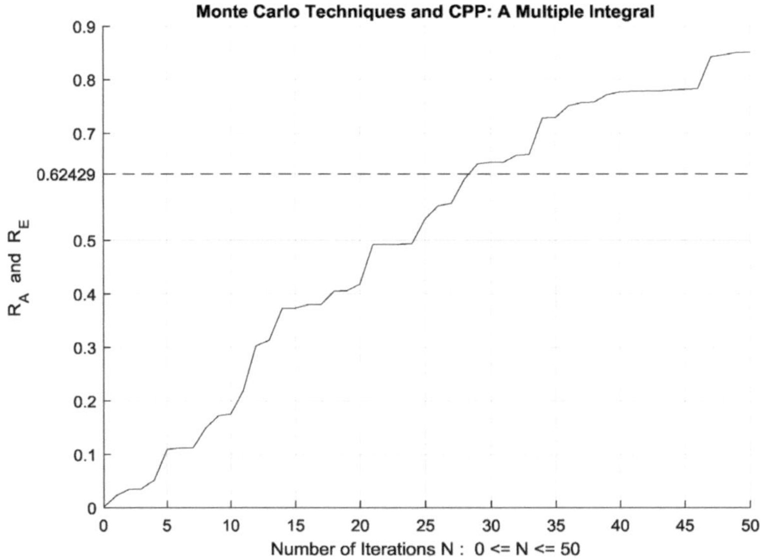

Figure 3.
The increasing convergence of the Monte Carlo method up to N = 50 iterations.

$$\Leftrightarrow \int_0^{4/3}\int_0^{4/3}\int_0^{4/3}\int_0^{4/3} xyzw.dxdydzdw \cong \frac{[(4/3-0)\times(4/3-0)\times(4/3-0)\times(4/3-0)]}{N}\sum_{j=1}^{N} x_j y_j z_j w_j$$
$$= \frac{256/81}{N}\sum_{j=1}^{N} x_j y_j z_j w_j = R_A$$

with $1 \leq N \leq N_C$ after applying Monte Carlo method.

Furthermore, the four figures (**Figures 3–6**) illustrate and prove the increasing convergence of Monte Carlo simulation and technique to the exact result

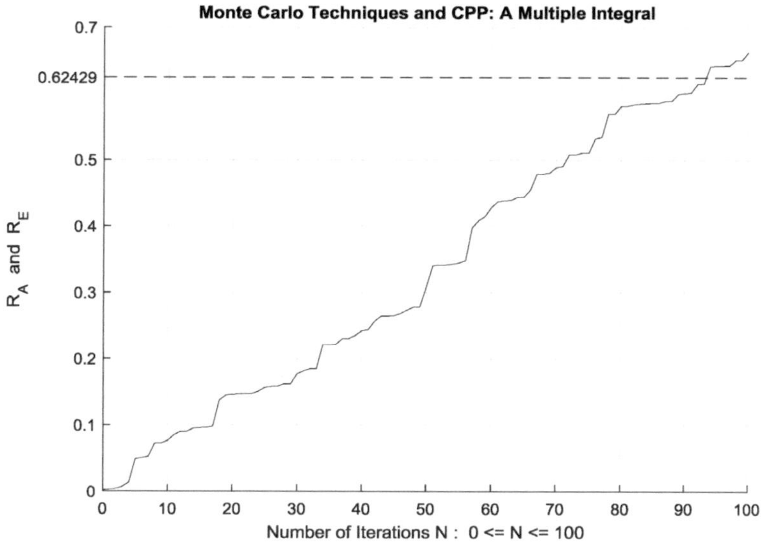

Figure 4.
The increasing convergence of the Monte Carlo method up to N = 100 iterations.

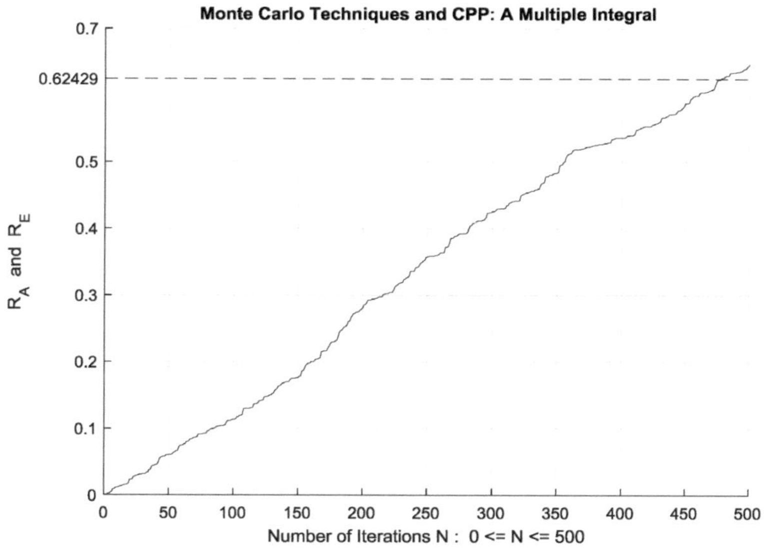

Figure 5.
The increasing convergence of the Monte Carlo method up to N = 500 iterations.

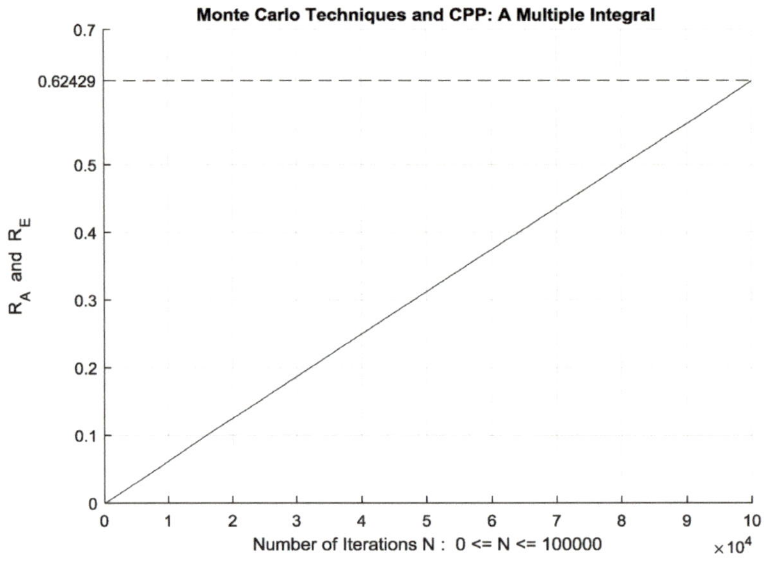

Figure 6.
The increasing convergence of the Monte Carlo method up to N = *100,000 iterations.*

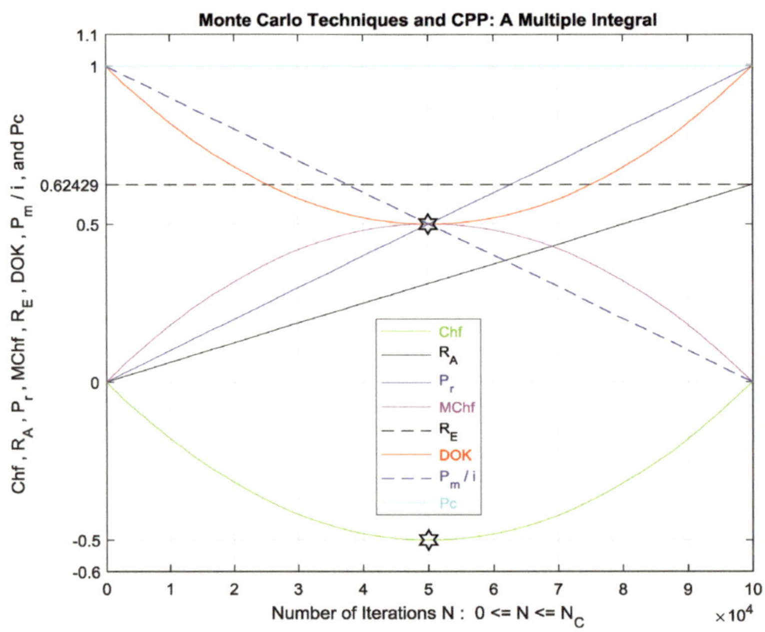

Figure 7.
The CPP *parameters and the Monte Carlo method for a multiple integral.*

$R_E = 0.62429507696997411...$ for N = 50, 100, 500, and $N = N_C = 100,000$ iterations. Consequently, we have $\lim_{N\to+\infty} P_r(N) = \lim_{N\to+\infty}\left\{1-\left|\frac{R_E-R_A(N)}{R_E}\right|\right\} = 1-\left|\frac{R_E-R_E}{R_E}\right| = 1-0 = 1$ which is equal to the probability of convergence of Monte Carlo technique as $N \to +\infty$.

Moreover, **Figure 7** shows undoubtedly and graphically the relation of all the parameters of the complex probability paradigm $(Chf, R_A, P_r, MChf, R_E, DOK, P_m/i, Pc)$ to the Monte Carlo technique after applying *CPP* to this four-dimensional integral.

7.2 The discrete random case: the matching birthday problem

An interesting problem that can be solved using simulation is *the famous birthday problem*. Suppose that in a room of n persons, each of the 365 days of the year (not a leap year) is equally likely to be someone's birthday. It can be proved from the theory of probability and contrary to intuition that only 23 persons need to be present for the probability to be better than fifty-fifty that at least two of them will have the same birthday.

Many people are interested in checking the theoretical proof of this statement, so we will demonstrate it briefly before doing the problem simulation. After someone is asked about his or her birthday, the probability that the next person asked will not have the same birthday is 364/365. The probability that the third person's birthday will not match those of the first two people are 363/365. It is well-known that the probability of two independent and successive events happening is the product of the probability of the separate events. In general, the probability that the nth person asked will have a birthday different from that of anyone already asked is

$$P(\text{all } n \text{ birthdays are different}) = \left(\frac{365}{365}\right) \times \left(\frac{364}{365}\right) \times \left(\frac{363}{365}\right) \times \dots \times \left(\frac{365-(n-1)}{365}\right)$$

The probability that the nth person asked will provide a match is 1 minus this value:

$$P(\text{matching birthdays}) =$$

$$1 - \left(\frac{365}{365}\right) \times \left(\frac{364}{365}\right) \times \left(\frac{363}{365}\right) \times \dots \times \left(\frac{365-(n-1)}{365}\right)$$
$$= 1 - \frac{(365) \times (364) \times (363) \times \dots \times [365-(n-1)]}{365^n} = R_E$$

which shows that with 23 persons, the chances are 50.7%; with 55 persons, the chances are 98.6% or almost theoretically certain that at least two out of 55 people will have the same birthday. The table gives the theoretical probabilities of matching birthdays for a selected number of people n (**Table 1**).

Without using the probability theory, we can write a routine that uses the random number generator to compute the approximate chances for groups of n persons. Obviously, what is needed here is to choose n random integers from the set of integers $\{1, 2, 3, \dots, 365\}$ and to check whether there is a match. When we repeat this experiment a large number of times, we can calculate afterward the probability of at least one match in any gathering of n persons. Note that if $n \geq 366$, then $P(\text{matching birthdays}) = 1$ by the famous pigeonhole principle.

Furthermore, the four figures (**Figures 8–11**) illustrate and prove the increasing convergence of Monte Carlo simulation and technique to the exact result $R_E = 0.706316242719\dots$ for $n = 30$ and for $N = 50, 100, 500$, and $N = N_C = 750,000$ iterations. Consequently, we have $\lim_{N\to+\infty} P_r(N) = \lim_{N\to+\infty}\left\{1 - \left|\frac{R_E - R_A(N)}{R_E}\right|\right\} = 1 - \left|\frac{R_E - R_E}{R_E}\right| = 1 - 0 = 1$ which is equal to the probability of convergence of Monte Carlo technique as $N \to +\infty$.

Moreover, **Figure 12** shows undoubtedly and graphically the relation of all the parameters of the complex probability paradigm ($Chf, R_A, P_r, MChf, R_E, DOK, P_m/i, Pc$) to the Monte Carlo technique after applying *CPP* to this problem of matching birthday.

Number of people n	Theoretical probability = R_E
$n = 5$	$P = 0.027135573700$
$n = 10$	$P = 0.116948177711$
$n = 15$	$P = 0.252901319764$
$n = 20$	$P = 0.411438383581$
$n = 22$	$P = 0.475695307663$
$n = 23$	$P = 0.507297234324$
$n = 25$	$P = 0.568699703969$
$n = 30$	$P = 0.706316242719$
$n = 35$	$P = 0.814383238875$
$n = 40$	$P = 0.891231809818$
$n = 45$	$P = 0.940975899466$
$n = 50$	$P = 0.970373579578$
$n = 55$	$P = 0.986262288816$
$n = 100$	$P = 0.999999692751$
$n = 133$	$P = 0.999999999999$
$n = 365$	$P = 1.000000000000$

Table 1.
Some theoretical probabilities of matching birthdays for n people where $1 \le n \le 365$.

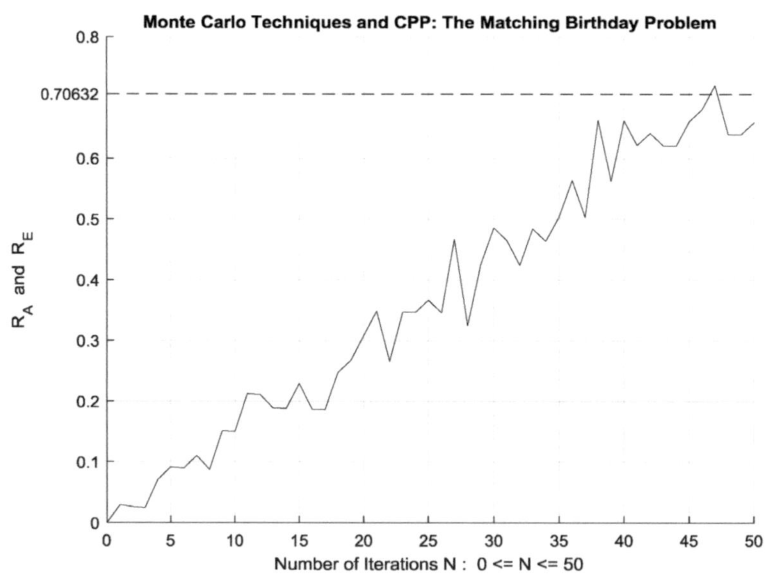

Figure 8.
The increasing convergence of the Monte Carlo method up to N = 50 *iterations.*

7.2.1 The cubes of complex probability

In **Figure 13** and in the first cube, the simulation of *Chf* and *DOK* as functions of the iterations N and of each other is executed for the problem of matching birthday. If we project $Pc^2(N) = DOK(N) - Chf(N) = 1 = Pc(N)$ on the plane $N = 0$ iterations,

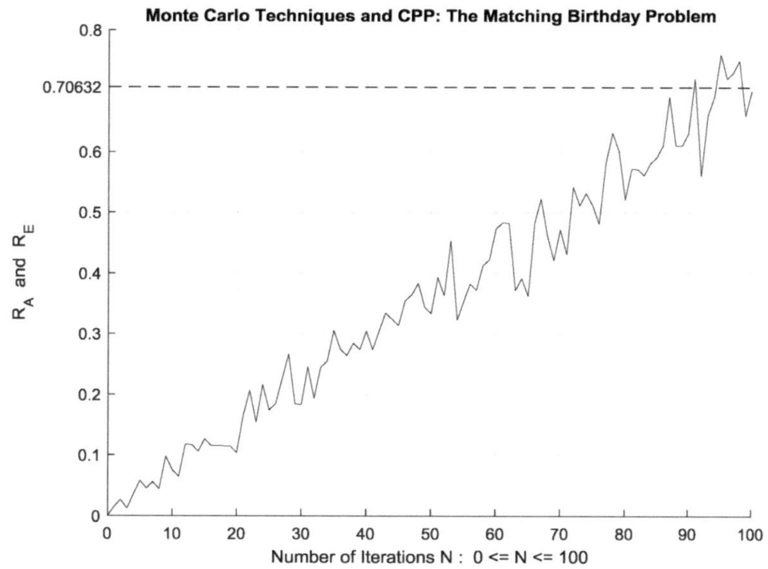

Figure 9.
The increasing convergence of the Monte Carlo method up to N = *100 iterations.*

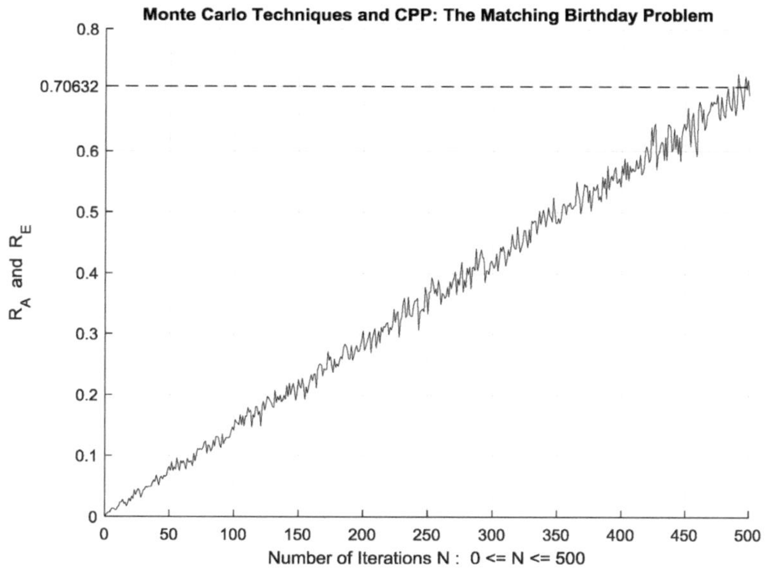

Figure 10.
The increasing convergence of the Monte Carlo method up to N = *500 iterations.*

we will get the line in cyan. The starting point of this line is point J ($DOK = 1$, $Chf = 0$) when $N = 0$ iterations, and then the line gets to point ($DOK = 0.5$, $Chf = -0.5$) when $N = 375{,}000$ iterations and joins finally and again point J ($DOK = 1$, $Chf = 0$) when $N = N_C = 750{,}000$ iterations. The graphs of $Chf(N)$ (pink, green, blue) in different planes and $DOK(N)$ (red) represent the other curves. We can notice that point K ($DOK = 0.5$, $Chf = -0.5$, $N = 375{,}000$ iterations) is the minimum of all these curves. We can notice also that point L has the coordinates ($DOK = 1$, $Chf = 0$, $N = N_C = 750{,}000$ iterations). Additionally, the three points J, K, and L correspond to the same points that exist in **Figure 12**.

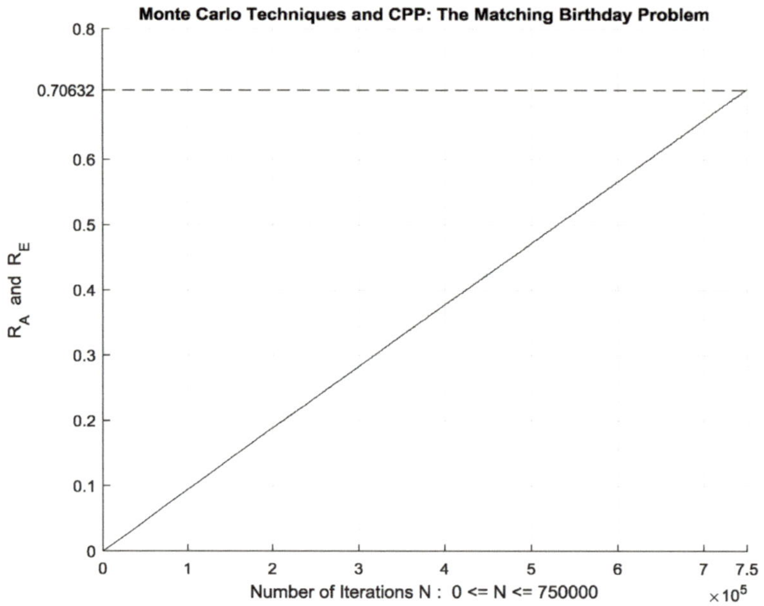

Figure 11.
The increasing convergence of the Monte Carlo method up to N = *750,000 iterations.*

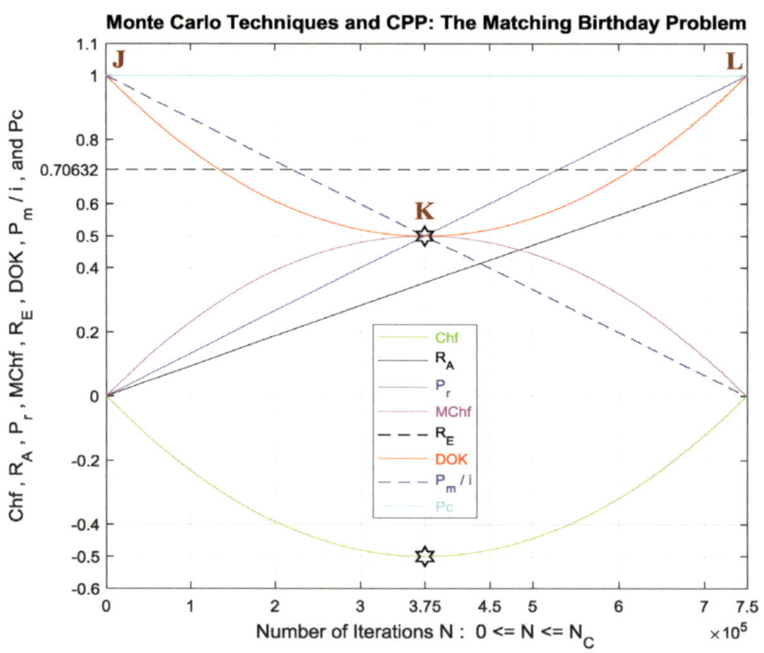

Figure 12.
The CPP *parameters and the Monte Carlo techniques for the matching birthday problem.*

In **Figure 14** and in the second cube, we simulate the probability of convergence $P_r(N)$ and its complementary real probability of divergence $P_m(N)/i$ as functions of the iterations N for the problem of matching birthday. If we project $Pc^2(N) = P_r(N) + P_m(N)/i = 1 = Pc(N)$ on the plane $N = 0$ iterations, we will get the line in cyan. The starting point of this line is point $(P_r = 0, P_m/i = 1)$, and the final point is

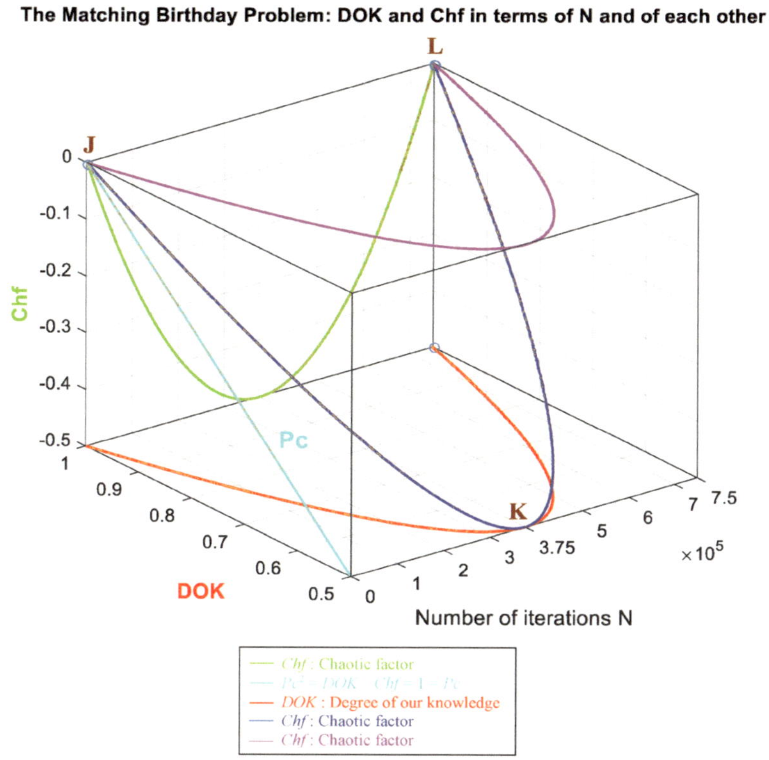

Figure 13.
Chf *and* DOK *in terms of each other and of* N *for the problem of matching birthday.*

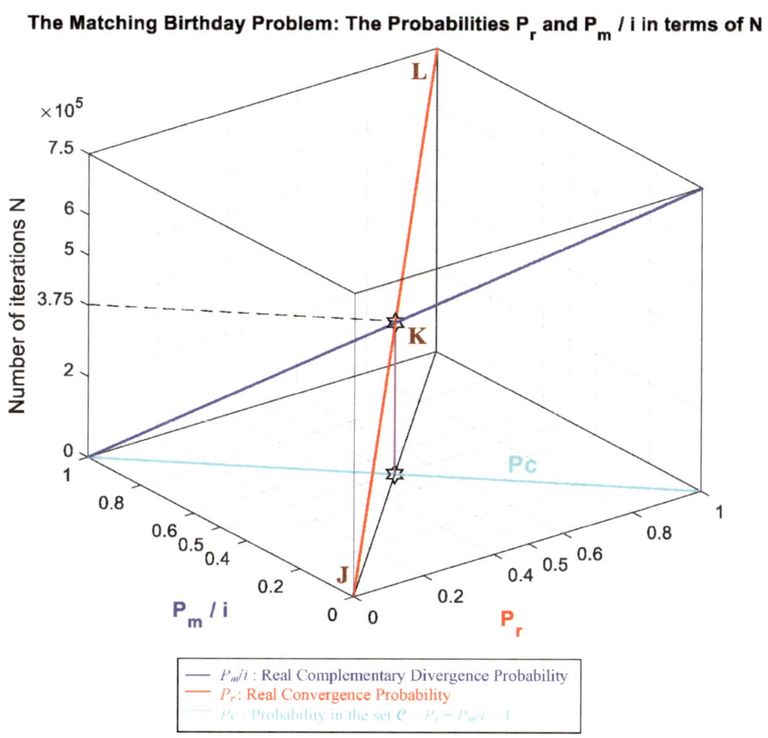

Figure 14.
P_m/i *and* P_r *in terms of each other and of* N *for the problem of matching birthday.*

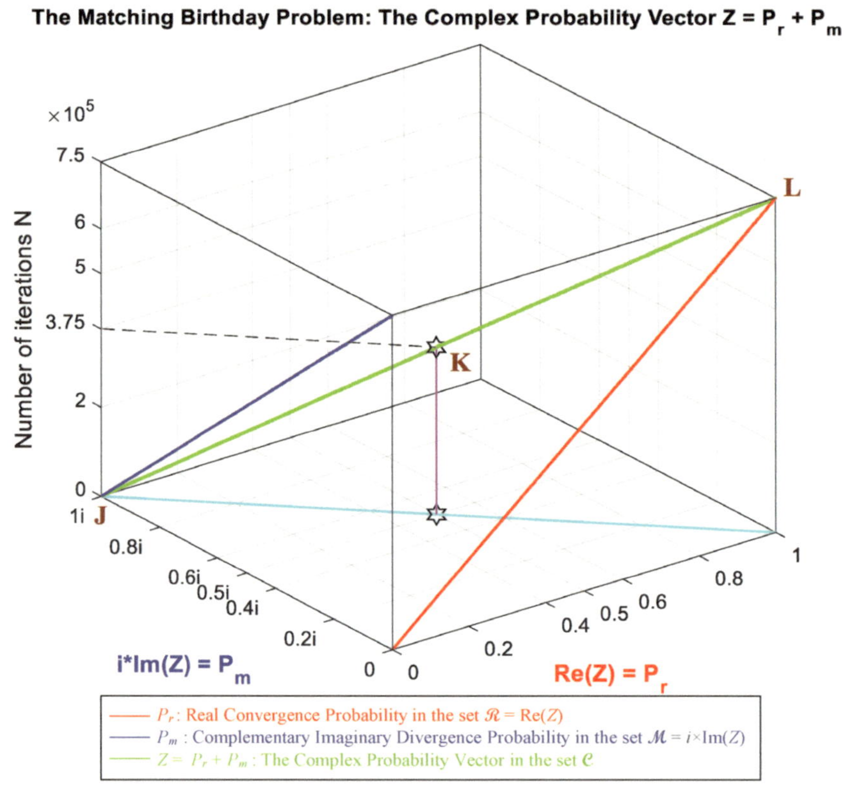

Figure 15.
The vector of complex probability Z *in terms of* N *for the problem of matching birthday.*

point ($P_r = 1, P_m/i = 0$). The graph of $P_r(N)$ in the plane $P_r(N) = P_m(N)/i$ is represented by the red curve. The starting point of this graph is point J ($P_r = 0, P_m/i = 1, N = 0$ iterations), and then it gets to point K ($P_r = 0.5, P_m/i = 0.5, N = 375{,}000$ iterations) and joins finally point L ($P_r = 1, P_m/i = 0, N = N_C = 750{,}000$ iterations). The graph of $P_m(N)/i$ in the plane $P_r(N) + P_m(N)/i = 1$ is represented by the blue curve. We can notice how much point K is important and which is the intersection of the blue and red graphs when $P_r(N) = P_m(N)/i = 0.5$ at $N = 375{,}000$ iterations. Additionally, the three points J, K, and L correspond to the same points that exist in **Figure 12**.

In **Figure 15** and in the third cube, we simulate the vector of complex probabilities $Z(N)$ in $\mathcal{C}$ as a function of the real probability of convergence $P_r(N) = \text{Re}(Z)$ in $\mathcal{R}$ and of its complementary imaginary probability of divergence $P_m(N) = i \times \text{Im}(Z)$ in $\mathcal{M}$, and as a function of the iterations N for the problem of matching birthday. The graph of $P_r(N)$ in the plane $P_m(N) = 0$ is represented by the red curve, and the graph of $P_m(N)$ in the plane $P_r(N) = 0$ is represented by the blue curve. The graph of the vector of complex probabilities $Z(N) = P_r(N) + P_m(N) = \text{Re}(Z) + i \times \text{Im}(Z)$ in the plane $P_r(N) = iP_m(N) + 1$ is represented by the green curve. The graph of $Z(N)$ has point J ($P_r = 0, P_m = i, N = 0$ iterations) as the starting point and point L ($P_r = 1, P_m = 0, N = N_C = 750{,}000$ iterations) as the end point. If we project $Z(N)$ curve on the plane of complex probabilities whose equation is $N = 0$ iterations, we get the line in cyan which is $P_r(0) = iP_m(0) + 1$. This projected line has point J ($P_r = 0, P_m = i, N = 0$ iterations) as the starting point and point ($P_r = 1, P_m = 0, N = 0$ iterations) as the end point. We can notice how much point K is important, and it corresponds to $P_r = 0.5$ and

$P_m = 0.5i$ when $N = 375{,}000$ iterations. Additionally, the three points J, K, and L correspond to the same points that exist in **Figure 12**.

8. Perspectives and conclusion

In the current chapter, the extended and original Kolmogorov model of eight axioms (*EKA*) was connected and applied to the random and classical Monte Carlo techniques. Thus, we have bonded Monte Carlo algorithms to the novel *CPP* paradigm. Accordingly, the paradigm of "complex probability" was more expanded beyond the scope of my 14 earlier studies on this topic.

Also, as it was proved and demonstrated in the original paradigm, when $N = 0$ (before the beginning of the random simulation) and when $N = N_C$ (after the convergence of Monte Carlo algorithm to the exact result), then the chaotic factor (*Chf* and *MChf*) is 0, and the degree of our knowledge (*DOK*) is 1 since the stochastic aspects and variations have either not commenced yet or they have terminated their job on the random phenomenon. During the course of the nondeterministic phenomenon ($N > 0$), we have $0 < MChf \leq 0.5$, $0.5 \leq DOK < 1$, and $-0.5 \leq Chf < 0$, and it can be noticed that throughout this entire process, we have continually and incessantly $Pc^2 = DOK - Chf = DOK + MChf = 1 = Pc$, which means that the simulation which seemed to be random and nondeterministic in the set $\mathcal{R}$ is now deterministic and certain in the set $\mathcal{C} = \mathcal{R} + \mathcal{M}$, and this after adding the contributions of $\mathcal{M}$ to the experiment happening in $\mathcal{R}$ and thus after removing and subtracting the chaotic factor from the degree of our knowledge. Additionally, the probabilities of convergence and divergence of the random Monte Carlo procedure that correspond to each iteration cycle *N* have been determined in the three sets of probabilities which are $\mathcal{C}$, $\mathcal{M}$, and $\mathcal{R}$ by Pc, P_m, and P_r, respectively. Subsequently, at each instance of *N*, the novel Monte Carlo techniques and *CPP* parameters *DOK*, *Chf*, *MChf*, R_E, R_A, P_r, P_m, P_m/i, Pc, and *Z* are perfectly and surely predicted in the set of complex probabilities $\mathcal{C}$ with Pc kept as equal to 1 continuously and forever. Also, referring to all these shown simulations and obtained graphs all over the entire chapter, we can visualize and quantify both the system chaos and stochastic influences and aspects (expressed by *Chf* and *MChf*) and the certain knowledge (expressed by *DOK* and *Pc*) of Monte Carlo algorithms. This is definitely very wonderful, fruitful, and fascinating and demonstrates once again the advantages of extending the five axioms of probability of Kolmogorov and thus the benefits and novelty of this original theory in applied mathematics and prognostics that can be called verily: "the complex probability paradigm."

Moreover, it is important to mention here that one essential and very well-known probability distribution was taken into consideration in the current chapter which is the uniform and discrete probability distribution as well as a specific generator of uniform random numbers, knowing that the original *CPP* model can be applied to any generator of uniform random numbers that exists in literature. This will yield certainly analogous results and conclusions and will confirm without any doubt the success of my innovative theory.

As a prospective and future challenges and research, we intend to more develop the novel conceived prognostic paradigm and to apply it to a diverse set of nondeterministic events like for other stochastic phenomena as in the classical theory of probability and in stochastic processes. Additionally, we will implement *CPP* to the first-order reliability method (FORM) in the field of prognostic in engineering and also to the problems of random walk which have huge consequences when applied to economics, to chemistry, to physics, and to pure and applied mathematics.

Nomenclature

$\mathcal{R}$	the events real set
$\mathcal{M}$	the events imaginary set
$\mathcal{C}$	the events complex set
i	the imaginary number with $i^2 = -1$ or $i = \sqrt{-1}$
EKA	extended Kolmogorov axioms
CPP	complex probability paradigm
P_{rob}	any event probability
P_r	the probability in the real set $\mathcal{R}$ = the probability of convergence in $\mathcal{R}$
P_m	the probability in the complementary imaginary set $\mathcal{M}$ that corresponds to the real probability set in $\mathcal{R}$ = the probability of divergence in $\mathcal{M}$
Pc	the probability in $\mathcal{R}$ of the event with its associated event in $\mathcal{M}$ = the probability in the set $\mathcal{C} = \mathcal{R} + \mathcal{M}$ of complex probabilities
R_E	the exact result of the random experiment
R_A	the approximate result of the random experiment
Z	complex probability number = complex random vector = sum of P_r and P_m
$DOK = \|Z\|^2$	the degree of our knowledge of the stochastic experiment or system, it is the square of the norm of Z
Chf	the chaotic factor of Z
MChf	the magnitude of the chaotic factor of Z
N	the number of iterations cycles = number of random vectors
N_C	the number of iterations cycles till the convergence of Monte Carlo method to R_E = the number of random vectors till convergence.

Author details

Abdo Abou Jaoude
Department of Mathematics and Statistics, Faculty of Natural and Applied Sciences, Notre Dame University - Louaize, Lebanon

*Address all correspondence to: abdoaj@idm.net.lb

References

[1] Abou Jaoude A, El-Tawil K, Kadry S. Prediction in complex dimension using Kolmogorov's set of axioms. Journal of Mathematics and Statistics, Science Publications. 2010;**6**(2):116-124

[2] Abou Jaoude A. The complex statistics paradigm and the law of large numbers. Journal of Mathematics and Statistics, Science Publications. 2013; **9**(4):289-304

[3] Abou Jaoude A. The theory of complex probability and the first order reliability method. Journal of Mathematics and Statistics, Science Publications. 2013;**9**(4):310-324

[4] Abou Jaoude A. Complex probability theory and prognostic. Journal of Mathematics and Statistics, Science Publications. 2014;**10**(1):1-24

[5] Abou Jaoude A. The complex probability paradigm and analytic linear prognostic for vehicle suspension systems. American Journal of Engineering and Applied Sciences, Science Publications. 2015;**8**(1):147-175

[6] Abou Jaoude A. The paradigm of complex probability and the Brownian motion. Systems Science and Control Engineering, Taylor and Francis Publishers. 2015;**3**(1):478-503

[7] Abou Jaoude A. The paradigm of complex probability and Chebyshev's inequality. Systems Science and Control Engineering, Taylor and Francis Publishers. 2016;**4**(1):99-137

[8] Abou Jaoude A. The paradigm of complex probability and analytic nonlinear prognostic for vehicle suspension systems. Systems Science and Control Engineering, Taylor and Francis Publishers. 2016;**4**(1):99-137

[9] Abou Jaoude A. The paradigm of complex probability and analytic linear prognostic for unburied petrochemical pipelines. Systems Science and Control Engineering, Taylor and Francis Publishers. 2017;**5**(1):178-214

[10] Abou Jaoude A. The paradigm of complex probability and Claude Shannon's information theory. Systems Science and Control Engineering, Taylor and Francis Publishers. 2017;**5**(1): 380-425

[11] Abou Jaoude A. The paradigm of complex probability and analytic nonlinear prognostic for unburied petrochemical pipelines. Systems Science and Control Engineering, Taylor and Francis Publishers. 2017;**5**(1):495-534

[12] Abou Jaoude A. The paradigm of complex probability and Ludwig Boltzmann's entropy. Systems Science and Control Engineering, Taylor and Francis Publishers. 2018;**6**(1):108-149

[13] Abou Jaoude A. The paradigm of complex probability and Monte Carlo methods. Systems Science and Control Engineering, Taylor and Francis Publishers. 2019;**7**(1):407-451

[14] Abou Jaoude A. Analytic prognostic in the linear damage case applied to buried petrochemical pipelines and the complex probability paradigm. In: Fault Detection, Diagnosis and Prognosis. London, UK: IntechOpen; 2020. DOI: 10.5772/intechopen.90157

[15] Abou Jaoude A. The Computer Simulation of Monté Carlo Methods and Random Phenomena. United Kingdom: Cambridge Scholars Publishing; 2019

[16] Abou Jaoude A. The Analysis of Selected Algorithms for the Stochastic Paradigm. United Kingdom: Cambridge Scholars Publishing; 2019

[17] Abou Jaoude A. Applied mathematics: Numerical methods and

algorithms for applied mathematicians [PhD thesis]. Spain: Bircham International University. 2004. Available from: http://www.bircham.edu

[18] Abou Jaoude A. Computer science: Computer simulation of Monté Carlo methods and random phenomena [PhD thesis]. Spain: Bircham International University. 2005. Available from: http://www.bircham.edu

[19] Abou Jaoude A. Applied statistics and probability: Analysis and algorithms for the statistical and stochastic paradigm [PhD thesis]. Spain: Bircham International University. 2007. Available from: http://www.bircham.edu

[20] Metropolis N. The Beginning of the Monte Carlo Method. Los Alamos Science (1987 Special Issue dedicated to Stanislaw Ulam); 1987. pp. 125-130

[21] Eckhardt R. Stan Ulam, John von Neumann, and the Monte Carlo Method. Los Alamos Science, Special Issue (15); 1987. pp. 131-137

[22] Mazhdrakov M, Benov D, Valkanov N. The Monte Carlo Method. Engineering Applications. Cambridge: ACMO Academic Press; 2018. p. 250. ISBN: 978-619-90684-3-4

[23] Peragine M. The Universal Mind: The Evolution of Machine Intelligence and Human Psychology. San Diego, CA: Xiphias Press; 2013. [Retrieved: 17 December 2018]

[24] McKean HP. Propagation of chaos for a class of non-linear parabolic equations. In: Lecture Series in Differential Equations, Session 7. Arlington, VA: Catholic University; 1967. pp. 41-57. Bibcode: 1966PNAS...56.1907M. DOI: 10.1073/pnas.56.6.1907. PMC: 220210. PMID: 16591437

[25] Herman K, Theodore HE. Estimation of particle transmission by random sampling. National Bureau of Standards: Applied Mathematics Series. 1951;**12**:27-30

[26] Turing AM. Computing machinery and intelligence. Mind. LIX. 1950;**238**: 433-460. DOI: 10.1093/mind/LIX.236.433

[27] Barricelli NA. Symbiogenetic evolution processes realized by artificial methods. Methodos. 1957:143-182

[28] Del Moral P. Feynman–Kac Formulae. Genealogical and Interacting Particle Approximations. Series: Probability and Applications. Berlin: Springer; 2004. p. 575

[29] Assaraf R, Caffarel M, Khelif A. Diffusion Monte Carlo methods with a fixed number of walkers. Physical Review E. 2000;**61**(4):4566-4575. Bibcode: 2000PhRvE...61.4566A. DOI: 10.1103/physreve.61.4566. Archived from the original (PDF) on 2014-11-07

[30] Caffarel M, Ceperley D, Kalos M. Comment on Feynman–Kac path-integral calculation of the ground-state energies of atoms. Physical Review Letters. 1993;**71**(13):2159. Bibcode: 1993PhRvL...71.2159C. DOI: 10.1103/physrevlett.71.2159. PMID: 10054598

[31] Hetherington JH. Observations on the statistical iteration of matrices. Physical Review A. 1984;**30**(2713): 2713-2719. Bibcode: 1984PhRvA...30.2713H. DOI: 10.1103/PhysRevA.30.2713

[32] Fermi E, Richtmyer RD. Note on Census-Taking in Monte Carlo Calculations (PDF). LAM. 805 (A). Declassified Report Los Alamos Archive; 1948

[33] Rosenbluth MN, Rosenbluth AW. Monte-Carlo calculations of the average extension of macromolecular chains. The Journal of Chemical Physics. 1955; **23**(2):356-359. Bibcode: 1955JChPh...23...356R. DOI: 10.1063/1.1741967

[34] Gordon NJ, Salmond DJ, Smith AFM. Novel approach to nonlinear/non-Gaussian Bayesian state estimation. Radar and Signal Processing, IEE Proceedings F. April 1993;**140**(2): 107-113. DOI: 10.1049/ip-f-2.1993.0015. ISSN 0956-375X

[35] Kitagawa G. Monte Carlo filter and smoother for non-Gaussian nonlinear state space models. Journal of Computational and Graphical Statistics. 1996;**5**(1):1-25. DOI: 10.2307/1390750. JSTOR 1390750

[36] Carvalho H, Del Moral P, Monin A, Salut G. Optimal non-linear filtering in GPS/INS integration. IEEE Transactions on Aerospace and Electronic Systems. 1997;**33**(3):835-850

[37] Del Moral P, Rigal G, Salut G. Nonlinear and Non-Gaussian Particle Filters Applied to Inertial Platform Repositioning. LAAS-CNRS, Toulouse, Research Report No. 92207, STCAN/DIGILOG-LAAS/CNRS Convention STCAN No. A.91.77.013. 1991. p. 94

[38] Crisan D, Gaines J, Lyons T. Convergence of a branching particle method to the solution of the Zakai. SIAM Journal on Applied Mathematics. 1998;**58**(5):1568-1590. DOI: 10.1137/s0036139996307371

[39] Crisan D, Lyons T. Nonlinear filtering and measure-valued processes. Probability Theory and Related Fields. 1997;**109**(2):217-244. DOI: 10.1007/s004400050131

Chapter 2

ANFIS TVA Power Plants Availability Modeling Development

Isa Qamber and Mohamed Al-Hamad

Abstract

In the present chapter, the evaluation of the Tennessee Valley Authority (TVA) Markov model transient behavior is derived and studied. It is focused on finding the models of the transient-state availability and unavailability of the four (TVA) models among using an adaptive neuro-fuzzy inference system (ANFIS). The developed ANFIS model for the TVA models is derived, and both availability and unavailability of the four TVA models are derived using the curve fitting technique, where each model of the transient availability of the three-state models of the TVA models is found. Each model is considered as a three-state model, and its equations obtained using the curve fitting technique are helping for the future availabilities and unavailabilities. The availability is a very important measure of performance for the availability of TVA power plants. The technique is used and applied on the four models in the present study to formulate and obtain the TVA models' results and are compared. In addition, the generation effects on the reliability investigation. The generation study evaluates the improvement in reliability over a time.

Keywords: transient availability, TVA, Markov model, ANFIS, curve fitting

1. Introduction

The Tennessee Valley Authority (TVA) is a business organization in the United States. TVA has been a key force for success in the Tennessee Valley since 1933 [1]. TVA supplies electricity to commercial customers and local energy supply companies. It serves approximately 10 million people in parts of seven southeastern states. TVA not only serves and invests its revenues in its electrical system but also provides flood control, navigation, and land management on the Tennessee River system and provides support to local energy companies.

Ren et al. [2] in their study went through the multi-microgrid system which is studied as a basic part of the intelligent network, and the radical system for several microgrids is the complete and standard system. The evaluation of the reliability of multi-microgrid systems is widely discussed to guarantee its reliability and steady-state operation. Balancing power between supply and demand is the key criterion for assessing the reliability of multiple microsystems, such as equipment shortages and inadequate distribution capacity in the network. In their research, few have developed a reliability model for partial failures and incomplete equipment repairs, and the transferability of distribution network (DN) is generally overlooked.
In their study, they summarize the multi-microgrid (MMG) radial system as a

performance sharing system and present a uniform modeling approach based on the Bayesian network for performance and reliability to fill research gaps. The operation of MMG radial systems is analyzed to create an abstract model to simplify the problem. The operational program is provided for the BN reliability assessment method. In addition, system modeling and BN parameter modeling are introduced. Finally, the MMG radial system composed of nine micronetworks (MNs) were examined and the variability of the system reliability indicator is analyzed in network-connected mode and in island mode.

Pham et al. [3] in their study highlight on the microgrid which mainly consists of distribution generators and energy storage systems, which are used to supply local loads. Distributed generators are mainly renewable energy sources. The aggregated system with numerous battery energy storage devices should be used to improve the reliability of the power supply from these renewable energy sources in the microgrid as a collective storage system battery power. The complete battery energy storage system is used to control the balance of source charging power so that microgrid can operate with high stability and reliability to supply electricity to a variety of customers. To demonstrate the importance of the complete battery energy storage system in microgrid, the reliability of its operation is examined in their study. Microgrid offers a systematic way to assess the reliability performance under various dynamic operational issues. An analytical method based on Markov models has been developed to assess the reliability of the entire energy storage system of the overall battery. In addition to the time-dependent failure rate, the voltage-dependent failure rate, and the power loss failure rate, important components, such as bilateral DC-to-DC converters, DC-to-AC converters, switching and protection devices, battery modules, and a battery charger/controller, are also designed and included in the reliability assessment. Depending on the dynamic operating problems of the microarrays with the full battery energy storage system and the photovoltaic (PV) generation systems, the overall battery energy storage system will be affected differently. The dynamic random operating conditions of the microgrid analyzed include the change in the load power, the intermittent and unstable operation of the PV sources, the modes of operation and distribution of the microarrays, and the cost/off-grid conditions. The complete battery energy storage system is used to control the balance of source charging power. The results of the simulation tests are presented and discussed to confirm that the operational reliability of the entire battery energy storage system in the microgrid strongly depends on its different dynamic operating strategies as well as activated voltage constraints.

Min et al. [4] in their paper highlight on the plan to switch to renewable energy, which recently has been a key element in the energy policy of the electricity system in South Korea. The renewable energy has raised questions about the reliability and flexibility of the electricity system. The researchers provide a research framework to assess the new policy in South Korea from different dimensions using three consecutive simulation models. The first-generation model in their study, which they derived from the best electricity generation mix, provides the overall cost of production and the environmental impact of the long-term capacity expansion plan. In addition, the researchers deal with other two simulation models to assess the reliability and flexibility of electrical systems, respectively. Also, they introduced the way to measure the results predictability of applying the framework in the new policy which shows that the policy does not guarantee a target level of reliability and increases costs and emissions. Achieving target system reliability requires additional reliability, and system flexibility is very sensitive to the type of capacity added.

Čepin, in his paper [5], published the modern electrical systems which introduce an increasing number of more dispersed and smaller energy sources, which are slowly replacing larger and more compact energy sources. The aim of the article is to examine the relocation of nuclear power plants to wind power plants to compare each scenario in terms of reliability of the electricity network. The author highlights on the load diagram and the variability of power plants. The load diagram power depends on environmental parameters over time. In addition, the article's updated method measures the power of the installation as a function of time instead of its nominal power. The first model, in the study, includes a dozen power plants including a nuclear power plant. This power plant will then be replaced by three wind turbines whose total power will be five times that of the nuclear power plant. Reliability is compared to pressure drop using actual weather data for a calendar year. The results show that a decrease in the reliability of the power system leads to a decrease in the reliability of the power system.

To determine the dynamic systems of a nuclear power plant, it is necessary to consider the effects of dynamic interactions, such as components, software, and operating processes. But at present, Lee et al. [6] in their study concentrate that there is no simple and easy-to-use tool to assess the availability of these dynamic systems. The method, for example the Markov chains, has a precise solution, but it is difficult to model the system. Using standard error trees, the reliability of a system with dynamic characteristics cannot be accurately estimated because error trees measure the reliability of a system configuration. The dynamic reliability graph with general gates (DRGGGs) allows intuitive modeling similar to actual system configuration, which can reduce the human error that occurs when modeling the target system. As the current dynamic reliability graph with general gates cannot assess the dynamic system in terms of reliability without repair, a new evaluation method enabling the availability of the dynamic repair system to be calculated is proposed through this study. The proposed method extends dynamic reliability graph with general gates by adding the repair condition to dynamic doors. A comparison of the method proposed by the Markov chains in terms of a simple validation model shows that the measured value converges toward the solution.

Sabouhi et al. [7] in their paper deal with the proper functioning of a power tool which depends on its subsystem and its components. Due to the current financial constraints in the energy sector, power plant operators face a wide range of challenges when dealing with the maintenance schedule and asset management practices. Knowing the roles and judgment of the components on the overall performance of the plant will help to plan for smooth, safe, and economical operation. To determine the technical and economic decisions for the maintenance of power plant equipment, this study focuses on modeling the reliability of combined cycle power plants (CCPPs). Reliability models are first developed for gas turbine power plants (GTPPs) and steam turbine power plants (STPPs), which provide the data needed to assess the overall reliability of the CCPP from a system perspective. Reliability indices with the reliability of the abovementioned types of supply devices are recommended to identify the critical components of a plant, that is, those which have the most significant impact on the reliability and availability objectives of the system. By identifying essential system components, it is possible to determine effective maintenance strategies for power tool components so that the available resources are well designed and technically allocated.

Uncertain sources of intermittent generation and load demand in addition to the transmission line access is not a threat to the security of electricity grids. In Sharifzadeh et al.'s [8] study, to address these uncertainties, an optimal stochastic energy flow is recommended when examining security constraints. A site generation method is also presented to assess the uncertainty of wind generations and load requests when evaluating their connections. In the proposed model, uncertainty is addressed through a combination of common decisions to be the best decision. The efficiency of the proposed model is demonstrated in the known 24 IEEE bus test system. The greater efficiency of the proposed model is shown numerically compared to four other definitive and stochastic methods for determining the reserve obtained and the "subsequent" conditions. In addition, the impact of the number of cases on the performance of the proposed model is assessed using a sensitivity analysis. Slower changes in the conditions created have also been shown to reflect correlations and can better capture the uncertain load behavior.

In recent years, there has been growing interest in developing maintenance activities for nuclear power plants in the form of risk-based models as studied by Mohammadhasani and Pirouzmand [9]. Maintenance design plays a key role in security likelihood assessment applications. The importance of these actions, especially when considering the effects of component degradation, is very clear for resolving controversial goals such as achieving the highest level of opportunity and reducing implementation costs of these actions, given the limitations of using standard safety assessment methods to test complex maintenance policies, as well as the difficulty of formulating the impact of component maintenance and degradation strategies using analysis of fault trees. Mohammadhasani and Pirouzmand [9] in their article present Markov's multi-time continuous approach to modeling three-component test and repair policies focusing on changing technical specifications such as test times. First, Markov models will be developed for three different test policies considering the effects of degradation. These models are then applied to analyze the availability of essential components of an emergency cooling system for the core of a VVER-1000/V446 nuclear power plant as a case study. In the first policy, the other components are not tested further. The test method is carried out in accordance with the test plan as originally recorded. In the second policy, additional testing should take place after the repair of the defective part and, in the third policy, the remaining redundant components should be subjected to extensive testing after the first detection.

Managing instant access to content servers (source servers) in a network in knowledge-centric networks has received little attention in Banerjee et al.'s [10] study. Banerjee et al. [10] considered a case in their study where content repositories in an information-centric network are temporarily unavailable, perhaps for reasons such as device malfunction, interruption of power outage, or denial of service attacks. Unlike the traditional host-centric Internet, users can still be served on information-centric networks, because content is stored on network nodes. The authors [10] begin their study by observing whether caches continue to operate using their native cache and expulsion policies after content repositories are unavailable, and over time the diversity of network content decreases. Therefore, the authors [10] recommend freezing network caches as soon as the custodians are not available. Next, the authors [10] provide a routing and search algorithm, content analysis under server availability, which uses breadcrumbs to efficiently find content stored on network nodes and to satisfy user requests when guards are not available. The authors [10] perform in-depth simulations of the real Internet topology in the Icarus simulator and show that content analysis on server accessibility can easily detect content stored on the network. The content of the server availability survey exceeds the shortest content demand on average by 56% and meets up to 98.5% of the total server demand.

2. Power plant model

The four TVA power plant models [11] are differing by nature of different numerical values for transition rates and are shown in **Table 1**. These transition rates are shown for the four TVA models represented by **Figure 1**, which are relevant to power plants operated by the TVA. Each TVA model is formed as illustrated in **Figure 2**, which is called Markov model. The general model for each of the TVA model can be represented by the following differential equations:

$$\frac{dP_{up}(t)}{dt} = -(a+b)P_{up}(t) + c\,P_{derated}(t) + d\,P_{down}(t) \tag{1}$$

$$\frac{dP_{derated}(t)}{dt} = a\,P_{up}(t) - (c+e)P_{derated}(t) + f\,P_{down}(t) \tag{2}$$

$$\frac{dP_{down}(t)}{dt} = b\,P_{up}(t) + e\,P_{derated}(t) - (d+f)P_{down}(t) \tag{3}$$

The three differential equations are solved using the MATLAB Simulink 7.10 package to find the transient probabilities of each model of the TVA, where the three states of each model are defined as follows:

Up-state: the system operates at full capacity.
Derated-state: the system operates at less than full capacity due to generators outages.
Down-state: the system has no power at all due to forced outage rate.

And the initial probabilities for each model are P1 (0) = 1, P2 (0) = 0, and P3 (0) = 0. The results are obtained for the four TVA models. The MATLAB Simulink 7.10 package is used to obtain the transient-state probabilities for the four TVA models and

Model	**Transition rates (per hour)**					
	a	**b**	**c**	**d**	**e**	**f**
1	0.0003	0.0010	0.0225	0.0350	0.0008	0.0004
2	0.0006	0.0050	0.0400	0.1000	0.0004	0.0004
3	0.0005	0.0002	0.0240	0.0430	0.0001	0.0001
4	0.0010	0.0006	0.0200	0.1000	0.0002	0.0020

Table 1.
The four TVA models' transition rate values [11].

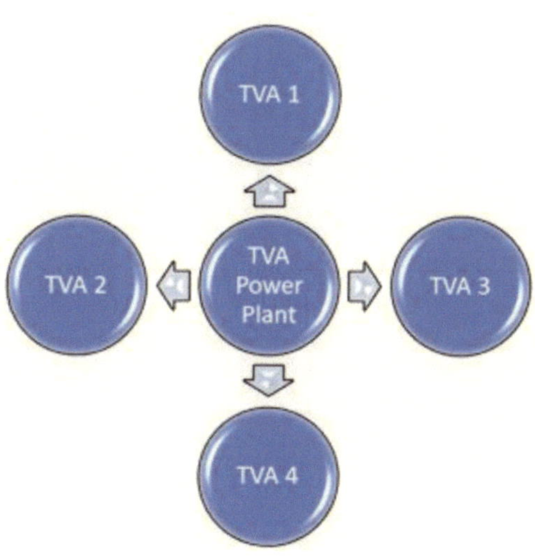

Figure 1.
TVA power plants' representation.

is reproduced by the fourth order Runge-Kutta method. Out of the obtained results, the availabilities and unavailabilities are obtained. The availability is the summation of the up-state and derated-state, where the unavailability is the down-state.

Table 2 summarizes the six transition rates of the four TVA models. **Figure 3a–d** illustrate the four TVA models with their transition rate values.

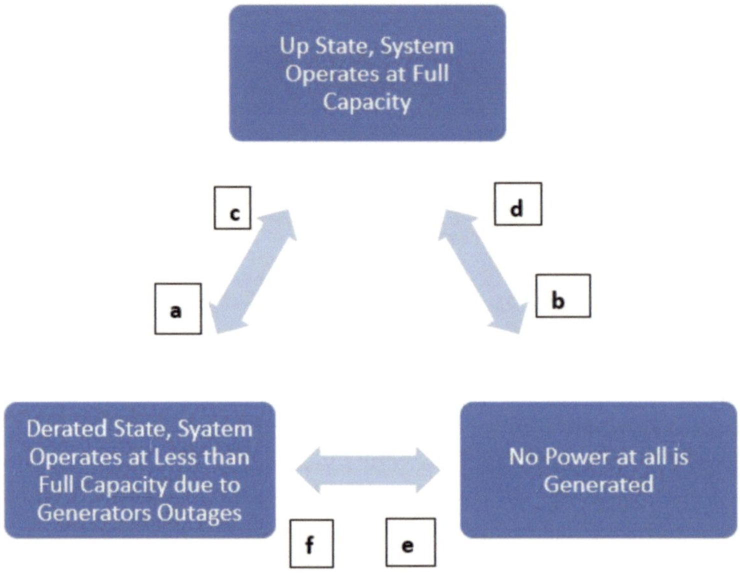

Figure 2.
TVA three-state Markov model.

Up-State, Derated-State, Transition Rate, a	Up-State, Down-State, Transition Rate, b
Derated-State, Up-State, Transition Rate, c	Down-State, Up- State, Transition Rate d
Derated-State, Down-State, Transition Rate, e	Down-State, Derated-State, Transition Rate, f

Table 2.
Transition rates of each TVA model.

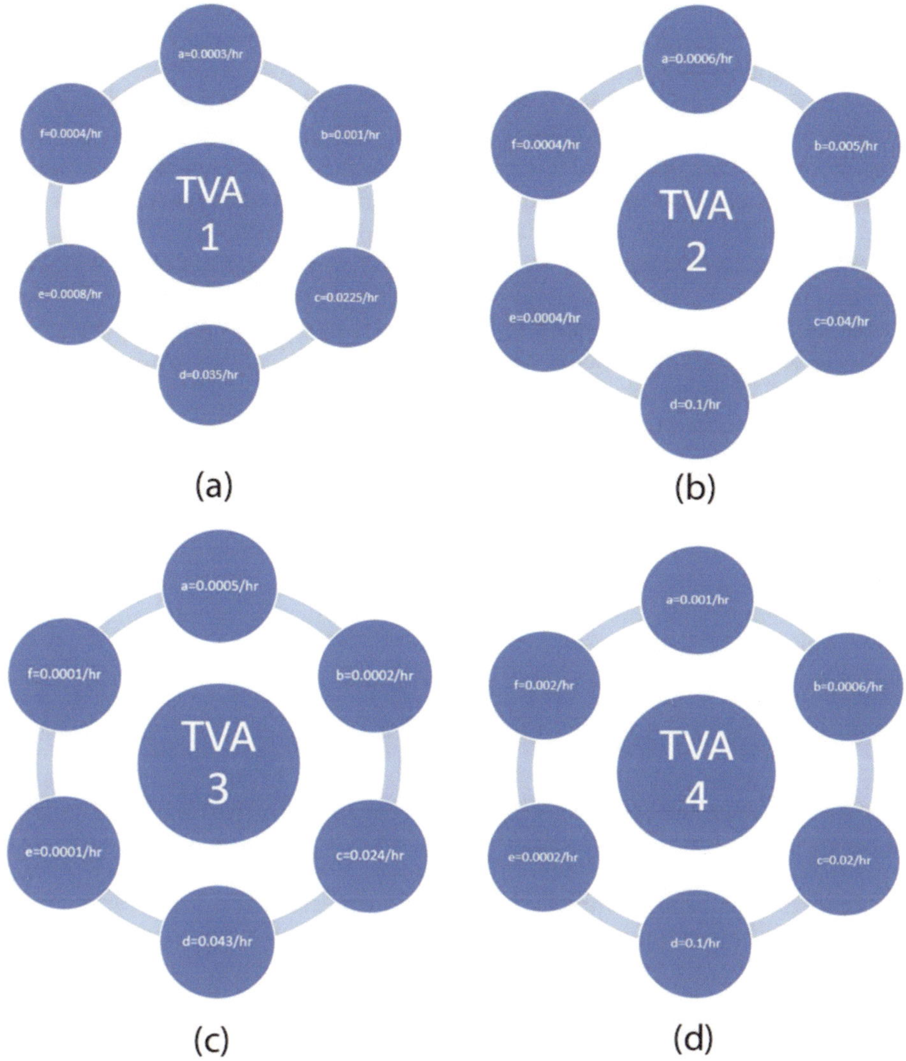

Figure 3.
Transition rate values for (a) TVA-1 model, (b) TVA-2 model, (c) TVA-3 model, and (d) TVA-4 model.

Al-Hamad and Qamber in their study [12] targeted a numerical evaluation of the nonlinear behavior of the Markov model and discussed it. Their study aims to obtain the transient probability between the states of four models recognized by the Tennessee Valley Authority (TVA). Three approaches are being studied to obtain the three transient probabilities after modeling. These techniques are Laplace transforms, curve fitting, and neuro-fuzzy. The MATLAB Simulink 7.10 package is used to obtain steady-state probabilities for the four VAT models while reproducing these solutions with Laplace transforms. For each model, a three-state model is considered, where its equations can be obtained using curve fitting and neuro-fuzzy methods. All the methods are used and applied in Al-Hamad and Qamber's study [12] and are used to design and obtain the TVA models. Al-Hamad and Qamber's study [12] proposes three technical approaches. These techniques increased the performance of the models in terms of execution time. These techniques are Laplace transformations, curve fitting, and neuro-fuzzy. The main objective of their study [12] is to temporarily reshape the transient-state probabilities for the four TVA models and find the best TVA models, even if any researcher has worked on modeling using curve fitting and other methods. TVA data are very useful for modeling a three-state model system. In their work the neuro-fuzzy was studied and used. It is found that it is the best and most useful to model and calculate the transient probabilities. In addition, the performance of the TVA models is clear and satisfying. As the way of calculating the transient probabilities throughout the operation period, each TVA model is likely to be trained at some point to develop fuzzy IF-THEN rules that help to determine the input and output variables for the functions of the TVA model. In Al-Hamad and Qamber's study [12], the combination of fuzzy logic and the neural network (becoming neuro-fuzzy) is a powerful means of designing intelligent systems and obtaining precise results. In their study [12], curved fitting modulation methods and neuro-fuzzy methods were adopted and applied to the four TVA models to calculate the probability of nonlinear state of the models. The performance of the two approaches was also evaluated to find the right and the accurate approach for the four TVA models. The results obtained from Al-Hamad and Qamber's study [9] were found to be much closer to the results of Laplace transforms, in particular, the results obtained using neuro-fuzzy. The study [12] shows a successful development of a reliable relationship between the probabilities of movement and time. Results obtained with both curve fitting and neuro-fuzzy were compared and studied. Comparing exposures to neuro-fuzzy offers better accuracy—as already mentioned—in predicting the probability of a moving state carried out in their study [12].

3. Applications and results

The neuro-fuzzy has a combination of advantages of the neural networks and fuzzy-logic, where the aim of the neuro-fuzzy is to combine collectively the benefits of both approaches. The neural networks have two main benefits. These two can learn nonlinear mapping of numerical data and the performing of parallel computation. It is very hard to understand the meaning of weights and the incorporation of prior knowledge into the system, which is usually impossible. Fuzzy logic uses human understanding of linguistic terms to form the knowledge of the system. This makes a close interaction between the system and human operator possible. In addition, neuro-fuzzy systems allow the incorporation of both numerical and linguistic data into the system, where it is also capable of extracting fuzzy knowledge from numerical data.

The ANFIS system applies the artificial neural network to find a suitable fuzzy inference system (FIS) structure and parameters. The fuzzy system with its structure identifies the considered fuzzy rules to obtain the targeted results. In addition, the considered architecture of the ANFIS structure has five layers as shown in **Figure 4**, which is the developed model in the study.

The developed general model for the neuro-fuzzy system is shown in **Figure 4**. **Figure 4** shows the developed five-layer connection for six inputs and two outputs. These five layers represent the neuro-fuzzy model, which is used to represent the TVA models and helps to obtain the availabilities and unavailabilities of each TVA model.

The proposed model using the neuro-fuzzy has six inputs as mentioned earlier and two outputs model as illustrated in **Figure 4**. In addition to the fourth order Runge-Kutta method, ANFIS technique applied in the present chapter to model the three-state probabilities of the four TVA models (fourth layer of **Figure 4**) which deals to the fifth layer helps to find the availability and unavailability of the model. The availability is reaching the steady-state availability of each TVA

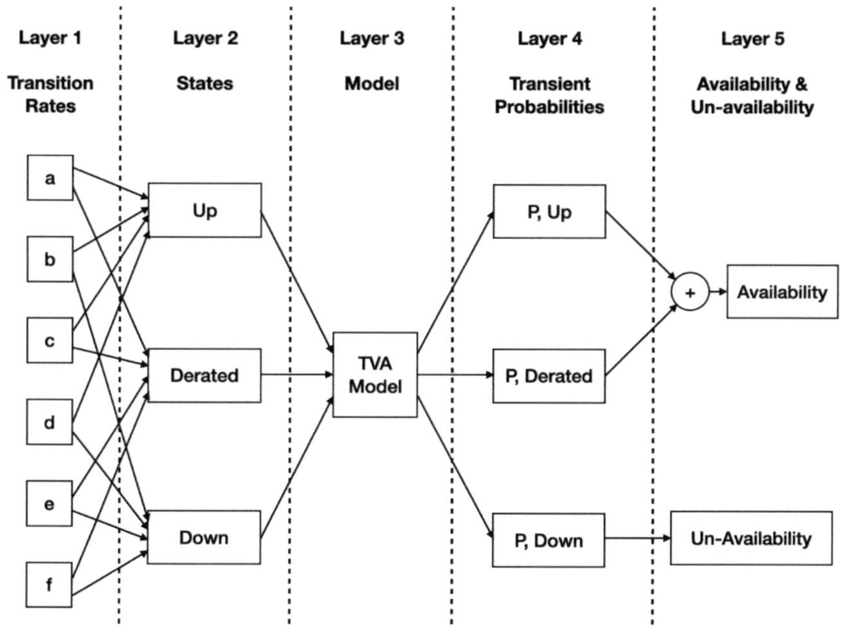

Figure 4.
ANFIS TVA developed model.

model, where the steady-state availability is defined as the steady-state probability of the system which is IN service. The steady-state probability that the TVA model is OUT of service is calculated as out of service due to failures. Any system can be interpreted as the long-run fraction of time spent in the failure state. The ANFIS and the curve fitting techniques are compared based on the availability and unavailability results. Through the obtained results, it is found that the results from both techniques are very close to each other. ANFIS and the curve fitting models are identical.

With reference to the TVA models, the fuzzy inference system (FIS) is applied. Six inputs are shown in layer (1), which are the transition rates between the three states model (**Figure 2**). The outputs of the model (layer (5)) results are the availability and unavailability of the TVA model. The second layer shows the three-states model. This process of converting an output fuzzy set for a solution variable into a single model is represented by the layer (3) which is TVA model. The results of the TVA model are shown in layer (4), which are the three probabilities P_{Up}, $P_{Derated}$, and P_{Down}.

The results obtained using the curve fitting for the four TVA models are shown in **Figures 5–8** as availability and unavailability. The curve fittings models' equations are obtained for the four TVAs' availability and unavailability as a function of time.

The availabilities as a function of time for the four TVA models are shown in **Figures 5a–8a**, respectively. The availability of each TVA model decreased gradually with time and then finally reached steady state (stable). The unavailability of each TVA model is illustrated in **Figures 5b–8b**, respectively. The unavailability of each TVA model increased gradually till it reached the steady state (stability).

When the generation reached the stability, the availabilities are 98.72%, 98.56%, 97.99%, and 95.26%, respectively, for the corresponding TVA models. Under the same conditions, the unavailabilities are 1.28%, 1.44%, 2.01%, and 4.74%, respectively, for the TVA models.

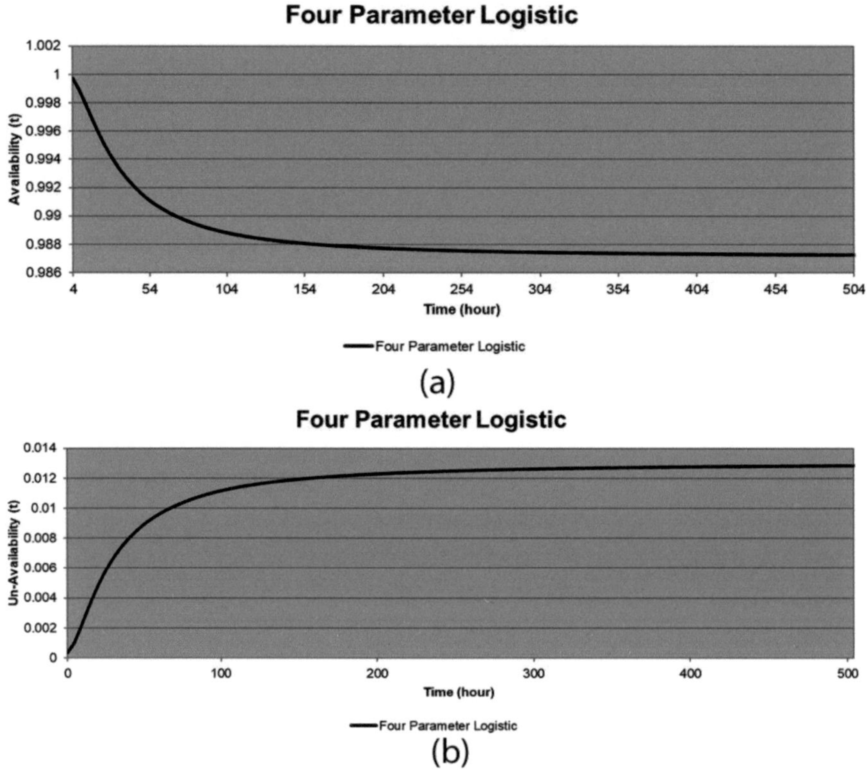

Figure 5.
(a) Availability model for TVA1 and (b) unavailability model for TVA1.

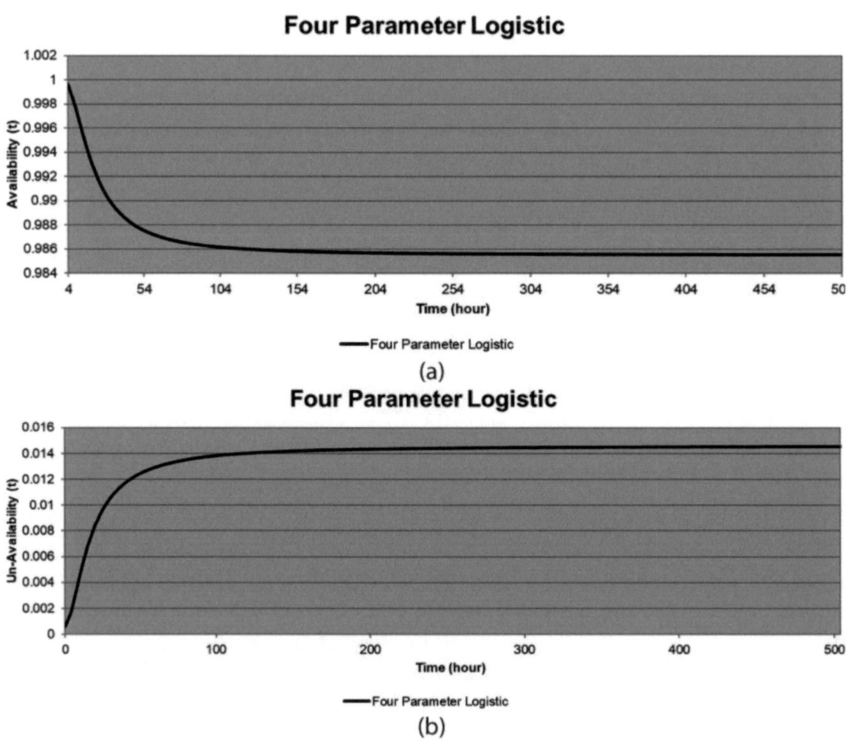

Figure 6.
(a) Availability model for TVA2 and (b) unavailability model for TVA2.

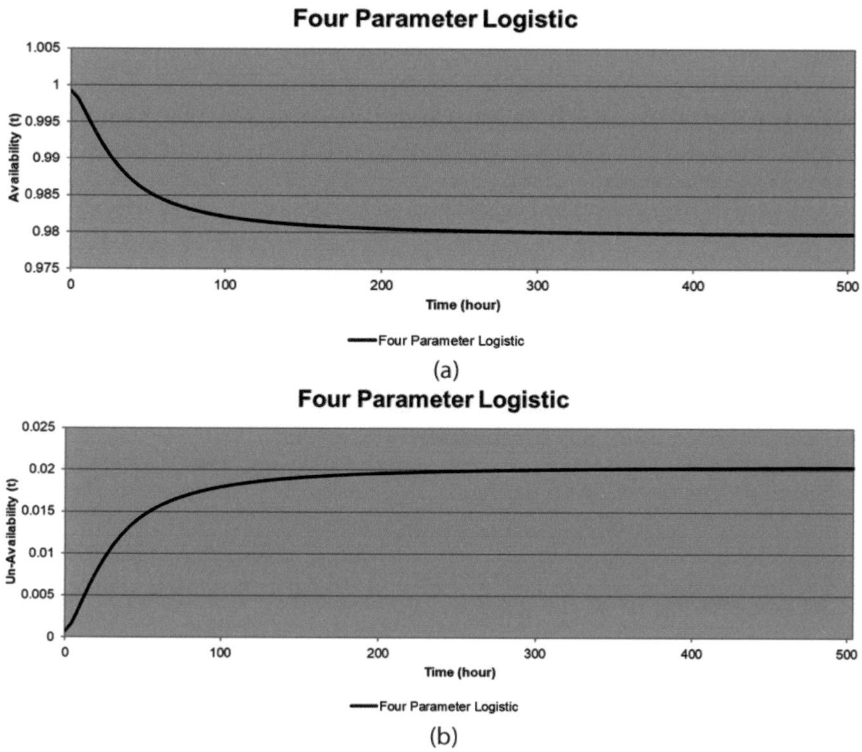

Figure 7.
(a) Availability model for TVA3 and (b) unavailability model for TVA3.

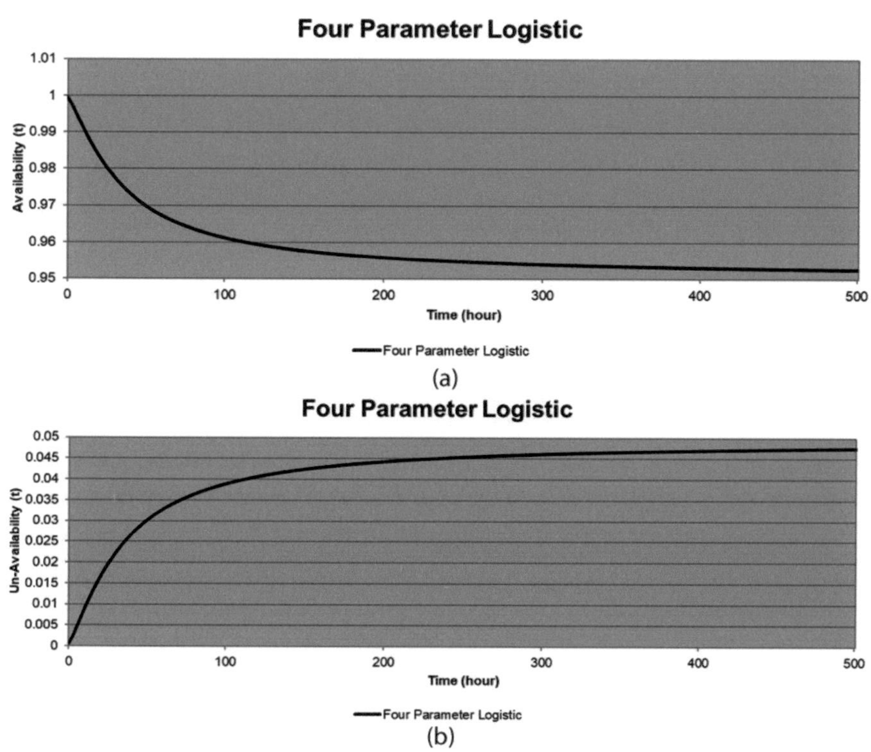

Figure 8.
(a) Availability model for TVA4 and (b) unavailability model for TVA4.

The curve fitting technique MyCurveFit (online curve fitting) is applied to find the model as shown in **Figures 5–8** as mentioned earlier. The results for the four TVA models are summarized in **Table 3**, showing the availabilities of each model, where **Table 4** shows the unavailabilities of the same models. The obtained availability equation using the curve fitting technique MyCurveFit is shown in Eq. (4) for the four TVA models.

$$Availability\,(time) = d + \frac{(a-d)}{1+\left[\dfrac{Time}{c}\right]^{b}} \tag{4}$$

The obtained unavailability equation using the curve fitting technique MyCurveFit is shown in Eq. (5) for the four TVA models.

$$Un - Availability\,(Time) = d + \frac{(a-d)}{1+\left[\dfrac{Time}{c}\right]^{b}} \tag{5}$$

Strong access to a complex system, such as a gas turbine, is linked to a part's reliability and maintenance policy. This policy affects not only the repair time of parts but also the reliability of parts, which affects the contamination and accessibility of the system. In any study, different methods are used for assessing the reliability and availability of gas turbines installed in a power plant. The investigated methods are based on concepts of system reliability, such as the development of the functional tree structure. The application of the failure mode and the analysis of the power

Availability model of TVA		Coefficients
TVA1	a	1.00014234742424
	b	1.646673111807
	c	32.8722769925419
	d	0.98712380612843
TVA2	a	1.00022688188532
	b	1.89339095988375
	c	20.6089357501677
	d	0.985496773685337
TVA3	a	0.999265231533308
	b	1.53764424728804
	c	29.218900622036
	d	0.97949890714863
TVA4	a	0.999505792270978
	b	1.26046003465127
	c	35.1534160245152
	d	0.950859080757418

Table 3.
Availabilities of the four TVA models.

Unavailability model of TVA		Coefficients
TVA1	a	0.000369658874527318
	b	1.47698177420345
	c	30.6343647596028
	d	0.0130413214169725
TVA2	a	0.000583437614778953
	b	1.64413595811526
	c	17.7140736131654
	d	0.0146075810962714
TVA3	a	0.000728795178208125
	b	1.52929468363619
	c	29.3840092626888
	d	0.0205824453759151
TVA4	a	0.000499149643278974
	b	1.26083143262919
	c	35.1382710744811
	d	0.0491382311673703

Table 4.
Unavailabilities of the four TVA models.

supply helps identify the critical components in improving the reliability of the system. The system and assessment of maintenance based on a historical fault database. This study focuses on the transition rates' changes (failure and repair rates) between the state which the model is passing through as observed and presented in this study. Implement trust-based maintenance concepts to implement the complex system maintenance policies of the generation system aimed at minimizing unexpected failure in critical components. The accessibility analysis shows different results for each TVA model, showing differences in the installation and operation of the system.

4. Conclusion

Various failure and repair transition rates have been taken in consideration in the present study. In the present study, the transition rates show the performance of the TVA models based on their variation and nature of each model. Different cases as obtained graphs show the availability and unavailability of the TVA models, where their models are produced by the curve fitting. The availability and unavailability values are obtained through a period of times. The unavailability increases until it is finally becoming stable.

The reliability of the availability and unavailability generation study seems important to obtain the means for the designer to apprehend the reliability for each design of TVA model. This means that the experience feedback is necessary. In the present study, the main part is to obtain the general TVA model transient availabilities and unavailabilities which has been modeled. This leads to the final target of the application study.

Acknowledgements

The authors would like to thank Ms. Aysha Isa Qamber for the great help during the preparation of this book chapter.

Conflict of interest

The authors declare no conflict of interest.

Author details

Isa Qamber[1*] and Mohamed Al-Hamad[2]

1 Former University of Bahrain, Bahrain Society of Engineers Member, Isa Town, Kingdom of Bahrain

2 GCC Interconnection Authority, Dammam, Kingdom of Saudi Arabia

*Address all correspondence to: i.s.qamber@gmail.com

References

[1] Tennessee Valley Authority. Available from: https://www.tva.com/About-TVA

[2] Yi R, Cui B, Feng Q, Yang D, Fan D, Sun B, et al. A reliability evaluation method for radial multi-microgrid systems considering distribution network transmission capacity. Computers & Industrial Engineering Journal. 2020;**139**:1-10. DOI: 10.1016/j.cie.2019.106145

[3] Pham TT, Kuo T, Bui DM. Reliability evaluation of an aggregate battery energy storage system in microgrids under dynamic operation. International Journal of Electrical Power and Energy Systems. 2020;**118**:1-22. DOI: 10.1016/j.ijepes.2019.105786

[4] Min D, Ryu J, Choi DG. Effects of the move towards renewables on the power system reliability and flexibility in South Korea. Energy Reports. 2020;**6**:406-417. DOI: 10.1016/j.egyr.2020.02.007

[5] Čepin M. Evaluation of the power system reliability if a nuclear power plant is replaced with wind power plants. Reliability Engineering and System Safety. 2019;**185**:455-464. DOI: 10.1016/j.ress.2019.01.010

[6] Lee EC, Shin SK, Seong PH. Evaluation of availability of nuclear power plant dynamic systems using extended dynamic reliability graph with general gates (DRGGG). Nuclear Engineering and Technology Journal. 2018;**51**:444-452. DOI: 10.1016/j.net.2018.10.009

[7] Sabouhi H, Abbaspour A, Fotuhi-Firuzabad M, Dehghanian P. Reliability modeling and availability analysis of combined cycle power plants. International Journal of Electrical Power & Energy Systems. 2016;**79**:108-119. DOI: 10.1016/j.ijepes.2016.01.007

[8] Sharifzadeh H, Amjady N, Zareipour H. Multi-period stochastic security-constrained OPF considering the uncertainty sources of wind power, load demand and equipment unavailability. Electric Power Systems Research. 2017;**146**:33-42. DOI: 10.1016/j.epsr.2017.01.011

[9] Mohammadhasani F, Pirouzmand A. Multi-state unavailability analysis of safety system redundant components with aging effect under surveillance testing. Progress in Nuclear Energy Journal. 2020;**126**:1-15. DOI: 10.1016/j.pnucene.2020.103415

[10] Banerjee A, Banerjee B, Seetharam A, Tellambura C. Content search and routing under custodian unavailability in information-centric networks. Computer Networks Journal. 2018;**141**:92-101. DOI: 10.1016/j.comnet.2018.05.015

[11] Biggerstaff BE, Jackson TM. The Markov process as a means of determining generating-unit state probabilities for use in spinning reserve applications. IEEE Transactions on Power Apparatus and Systems. 1969;**88**:423-430. DOI: 10.1109/TPAS.1969.292464

[12] Al-Hamad MY, Qamber IS. TVA generating unit modeling using MATLAB. Journal of the Association of Arab Universities for Basic and Applied Sciences. 2015;**17**:73-81. DOI: 10.1016/j.jaubas.2014.05.003

Chapter 3

A Layered Recurrent Neural Network for Imputing Air Pollutants Missing Data and Prediction of NO_2, O_3, PM_{10}, and $PM_{2.5}$

Hamza Turabieh, Alaa Sheta, Malik Braik and Elvira Kovač-Andrić

Abstract

To fulfill the national air quality standards, many countries have created emissions monitoring strategies on air quality. Nowadays, policymakers and air quality executives depend on scientific computation and prediction models to monitor that cause air pollution, especially in industrial cities. Air pollution is considered one of the primary problems that could cause many human health problems such as asthma, damage to lungs, and even death. In this study, we present investigated development forecasting models for air pollutant attributes including Particulate Matters ($PM_{2.5}$, PM_{10}), ground-level Ozone (O_3), and Nitrogen Oxides (NO_2). The dataset used was collected from Dubrovnik city, which is located in the east of Croatia. The collected data has missing values. Therefore, we suggested the use of a Layered Recurrent Neural Network (L-RNN) to impute the missing value(s) of air pollutant attributes then build forecasting models. We adopted four regression models to forecast air pollutant attributes, which are: Multiple Linear Regression (MLR), Decision Tree Regression (DTR), Artificial Neural Network (ANN) and L-RNN. The obtained results show that the proposed method enhances the overall performance of other forecasting models.

Keywords: imputing missing data, air pollutants, prediction, layered recurrent neural network

1. Introduction

Air quality monitoring and management have drawn much attention in recent years and attracted great attention from the public. Air pollution poses serious problems and infection for living organisms and environmental risks [1]. Harmful emission of industrial waste on air is one of the common environmental influences that disturb the air quality specifications and the national economy [2]. Significant publications have shown that air pollution has harmful effects on human health [3]. Air pollution affects the living organisms by producing impacts on cardiac, vascular, pulmonary, and neurological systems [4]. For example, air pollution in the City of New York causes the death of more than 3000 people and causes hospitalization of 200 persons [5]. It was found that many of these reported incidences were caused by the exposure to $PM_{2.5}$ and other pollutant attributes [6]. In 2010, it was estimated that ambient particulate matter (PM) caused 3.2 million premature deaths [7]. Moreover, several-analysis and research papers highlight that there is an exponential relationship between PM values and cardiovascular disease, and significant relation between NO_2 concentrations and cardiovascular disease [8, 9].

Air pollution arises from many sources such as vehicle fumes, agricultural, industrial, and natural sources like volcanoes [10]. Common air pollutants are classified into two groups: trace gases such as carbon monoxide (CO), nitrogen dioxide (NO_2), ground-level ozone (O_3), and sulfur dioxide (SO_2) or particulate matter ($PM_{2.5}$) or (PM_{10}) in aerodynamic diameter [11]. Tropospheric ground-level ozone (O_3) is a secondary factor that can damage human health and ecosystem [12–41]. O_3 concentration is one of the most serious oxidant factors that are harmful to human skin and lung tissues when inhaled [15, 16]. Several side effects impair pulmonary function and cause respiratory symptoms such as headache, weight loss, cough, shortness of breath, hoarseness, and pain while breathing [17]. Moreover, several epidemiological research studies focus on the relation between O_3 pollution and mortality [18].

1.1 Challenges

Air pollution monitoring and control is a major global challenge [19, 20]. To develop and train air quality prediction models, meteorological data for the investigated area should be collected and used. This data mostly consists of physical parameters that include temperature, dew point, wind direction, wind speed, cloud cover, cloud layer(s), ceiling height, visibility, current weather, amount of precipitation, and many more [21, 42]. These attributes greatly influence the concentration of pollutants in the area of interest.

Recently, cities are exposed to air pollutants either indoors or outdoors [22]. Several monitoring stations (i.e., sensors) are used to monitor the air quality by collecting data from different locations inside cities. These stations are used to collect data for gases or particulate matter in an accurate manner [23]. These sensors can be categorized as wired or wireless sensors. Wired sensors need great efforts for deployment and maintenance. Wired sensors can be easily breakdown due to several reasons (e.g., environment close to a volcano, where the hot gases and steams can damage a wired network easily [24, 43]). Wireless sensors still in an early stage. However, they show a great performance compared to wired sensors either in deployment or maintenance. Both types of sensors send the collected data to a central station for further processing. However, sometimes the process of collecting data suffers from different problems such as power failure, sensor fault, man-made error in measurements, and many others. **Figure 1** depicts the process of collecting data from different sensors. For example, if the gas sensor (i.e., O_3) does not work accurately, the collected data will not be complete and accurate. As a result, the air quality of the prediction model may not be accurate if the percentage of missing data is high.

Missing data cause serious problems for developing prediction models. The presence of missing data could severely reduce the quality of air quality prediction models. To solve this problem, we may either remove the missing data or imputing it. Removing the missing data may reduce the application performance [25], while imputing missing data may enhance the overall performance and without losing the collected data. Many methods exist to impute the missing data. Researchers either applied simple methods such as average value or complex methods such as machine

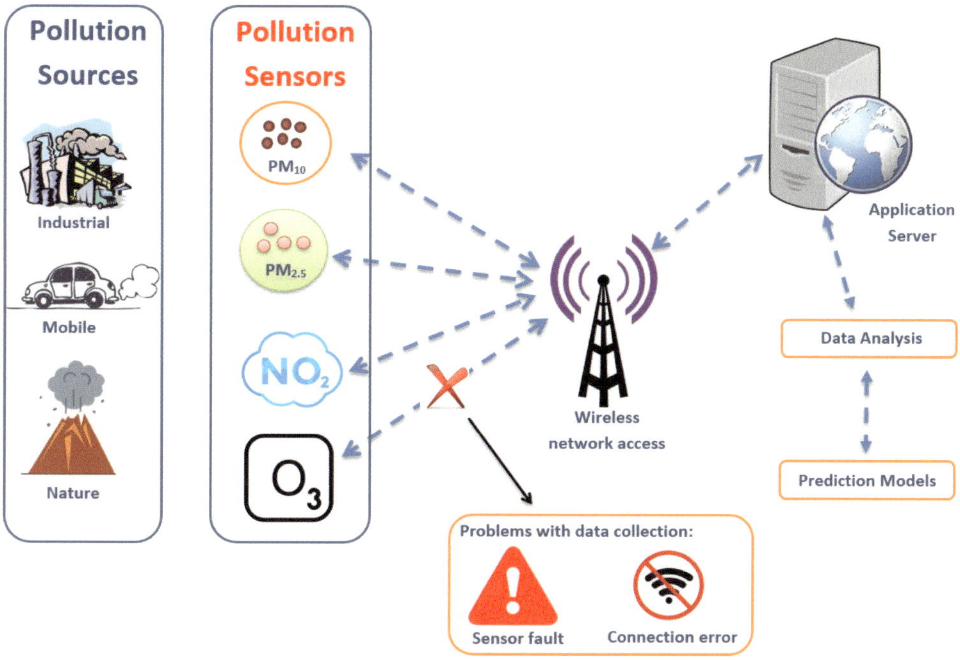

Figure 1.
The process of collecting data using air pollutants sensors.

learning methods to impute missing data [25]. Imputing missing values based on average is not accurate compared to the other one.

The main goal of this study is to propose a hybrid model that can predict the daily average of the concentration of air pollutants based on missing data imputation. The proposed model is a machine learning approach that can enhance the performance of monitoring systems of air pollution inside cities. Layered Recurrent Neural Network (L-RNN) [26] was successfully used to solve a variety of state-of-the-art applications such as detection of heart failure [27] and time-series data classification [28]. L-RNNs for missing data were explored earlier to handle the missing data problem [25, 29]. In this research, we first explore the use of L-RNN for imputing the missing data collected from Dubrovnik city that is in the east of Croatia. In the second step, we develop a series of models for predicting NO_2, O_3, PM_{10}, and $PM_{2.5}$ using the machine learning model (i.e., MLR, DTR, ANN, and L-RNN).

The rest of this chapter is organized as follows: In Section 2, the related works of air pollution as well as the literature of missing data is presented. Section 3 describes the proposed methodology. Section 4 presents predictive models using machine learning concepts. Section 5 presents the data collection process. The evaluation criteria used in this chapter are presented in Section 6. Section 7 presents the experimental setup used in this paper. Finally, a conclusion of the work is presented in Section 8.

2. Related works

2.1 Imputation vs. removing data

One of the most common problems in the process of developing prediction models is the Data Cleaning/Exploratory Analysis. This phase becomes a challenge when missing values are in presence. In general, there is no fundamental method to deal with missing data. Missing data problem occurs if no value(s) is assigned while collecting data. In general, the missing data are presented by different symbols such as ?, N/A, or –. There are two methods adopted in the literature to handle missing data. They are:

2.1.1 Deletion

Several researchers remove the missing data from the collected dataset if the percentage of the missing data is less than 5%. However, if the percentage of missing data is greater than 5%, the dataset should be examined carefully [30]. Many approaches have been investigated by researchers to solve the missing data problem. For example, the data list wise deletion method removes the missing data or incomplete data from the collected dataset. This method works fine if the percentage of missing data is very small and does not affect the overall accuracy [31]. The pairwise data deletion method keeps the missing data and tries to reduce the loss that occurs in the list wise deletion method. However, deleting missing values is an acceptable approach for some applications. Mary and Arockiam [32] investigated the missing data as a case study of air pollution. They proposed an ST-correlated proximate approach to impute the incomplete dataset for the air pollution system. The authors compared the obtained results of the proposed approach with different statistical methods. Sta [33] investigated the process of collecting data for modern urban cities. The author proposed a framework to cluster the collected data into three clusters: complete, ambiguous, and missing data. The author imputed the missing data and enhanced the overall performance of the proposed system. Xiaodong et al. [34] proposed a Hot Deck imputation approach that imputes the incomplete records (missing data) using the similarity between complete and incomplete data.

2.1.2 Imputation

In statistics, imputation is defined as the process of substituting missing data with swapped values. Unit imputation is used when we replace a single data point while the replacement of a component of a data point, is called, item imputation. Imputation is considered a successful solution to avoid difficulties associated with list wise deletion of missing values. Suhani et al. [35] proposed a machine learning approach based on the fuzzy kNN technique to impute the missing data for a selected case from the medical field. The authors ignore the missing data whose entropy value is less than a predetermined value and recover the incomplete data that are higher than the predetermined value based on a fuzzy kNN algorithm. Chen et al. [36] applied a machine learning approach based on a convolutional neural network to impute the missing data for a real medical dataset. The authors improve the overall performance after imputing missing data. Turabieh et al. [25] proposed a dynamic model based on deep learning neural networks for missing data imputation. The authors showed that the proposed model improves the overall performance of medical applications after imputing missing data.

2.2 Air pollution prediction

Air pollution is a serious problem that negatively affects human health, environment, and climate. Governments and organizations published several initiatives to reduce the concentrations of air pollutants, but high levels of concentrations of air pollutants still exist. As a result, monitoring the concentrations of air pollution is needed. Air monitoring consists of several steps; 1) Monitoring sites based on wired or wireless sensors, 2) collecting data that should be accurate and complete, 3) data analysis using predictive models based on machine learning to predict and analyze the collected data, and, 4) the final step is making decisions to reduce the concentrations of air pollution. This process should be performed correctly to ensure that the concentration of air pollution is under control.

Different types of machine learning methods have been used to predict the concentrations of air pollutant indicators by many researchers. For example, Perez and Gramsch [37] applied a feed-forward neural network to predict the concentration of $PM_{2.5}$ and PM_{10} in Santiago, Chile. The obtained accurate results show that the proposed approach enhances the prediction of $PM_{2.5}$ and PM_{10}. Lana et al. [38] employed regression models to predict several air pollutants such as CO, NO, NO_2, O_3 and PM_{10}) for the city of Madrid (Spain). The obtained results explore the importance of reducing air pollutants in the city of Madrid. Kamińska [39] employed an ensemble learning method based on random forests to model the relationship between the concentrations of air pollutants and nine variables describing meteorological conditions, temporal conditions, and traffic flow. The collected data was for 2 years 2015 and 2016 for WrocÅ,aw city. The data consists of hourly values of wind speed, wind direction, temperature, air pressure and relative humidity, temporal variables, and traffic flow. The obtained results show that the season plays a vital role in the overall performance. Kamińska [40] proposed a probabilistic forecasting method to predict the concentrations of NO_2. The dataset represents the hourly values of the concentration of NO_2 wind speed, and traffic flow for the main intersection in Wrocław city. The obtained results show that wind speed plays a vital factor in the concentration of NO_2.

Shang et al. [41] employed a novel prediction method that hybridized the regression tree (CART) and ensemble extreme learning machine (EELM) methods to predict the hourly concentration of $PM_{2.5}$ air pollutant. The training dataset used in this research obtained from the meteorological data of Yancheng urban area, while the testing data (i.e., the air pollutant concentration) obtained from the City Monitoring Centre. The obtained results demonstrate the effectiveness of the proposed method to predict $PM_{2.5}$. A hybrid framework based on three different machine learning methods (i.e., genetic algorithm [GA], random forests [RFs], and backpropagation neural networks [BPNN]) proposed by Dotse et al. [44]. The proposed hybrid approach is used to predict daily PM_{10} in Brunei Darussalam. Sun and Sun [45] proposed a hybrid model to predict $PM_{2.5}$ in Baoding city in China, where a combination of three machine learning methods (i.e., principal component analysis [PCA], least squares support vector machine [LSSVM], and cuckoo search [CS]). The obtained results show that the PCA algorithm works as a feature selection algorithm that reduces the dimensionality of the input dataset while CS shows promising results to predict $PM_{2.5}$. The main shortfall of this work that is applicable for short-term $PM_{2.5}$ forecasting.

A dynamic fuzzy synthetic evaluation model for predicting the concentration of three air pollutants (i.e., of $PM_{2.5}$, PM_{10} and SO_2) in two cities from China have been proposed by Xu et al. [46]. The obtained results show that the proposed model can be employed to build a robust monitoring air quality system for early warning. A novel hybrid model based on extreme learning machine (ELM) is employed to predict the concentration level of PM_{10} for Beijing and Harbin cities in China by Luo et al. [47]. Aznarte [48] proposed an ELM approach that is optimized by cuckoo search (CS) to enhance the overall performance of ELM. A probabilistic forecasting approach is applied to predict NO_2 in Madrid city from Spain. Wang et al. [49] proposed a novel hybrid machine learning approach based on a decomposition method and extreme learning machine (ELM) that is optimized by differential evolution (DE) to predict air pollutants in Beijing and Shanghai cities from China. Kumar and Goyal [50] applied Multiple Linear Regression (MLR) and Principal Component Regression (PCR) methods to predict several air pollutants in Delhi city from India. A MultiLayer Perceptron (MLP) neural network is adopted to predict PM_{10} in Delhi city from India by Aly et al. [51]. The authors also applied two algorithms (i.e., Naïve Bayes [NB] and Support Vector Machine [SVM]) and the performance of MLP outperforms NB and SVM. Vibha and Satyendra [52] applied seven models of neural networks using Levenberg–Marquardt (LM) to predict the daily PM_{10} in two cities from India.

3. Methodology

The main purpose of this research is to evolve different machine learning methods to predict daily average air pollutant concentrations such as O_3, $PM_{2.5}$, PM_{10}, and NO_2 values given data with many missing values. The process consists of two phases: 1) imputing missing data based L-RNN model, and 2) development of predictive models using several machine learning algorithms which include LR, DTR, ANN, and L-RNN. Our proposed approach starts by collecting data from sensors. If the collected data suffer from missing data, an imputation process will start based on the L-RNN hat predicts the concentration of air pollutants. This process will be repeated until the collected data have no missing value(s). Once the collected dataset is complete, a machine learning model is selected to predict the daily average of air pollutant concentrations. The selected model is evaluated based on two evaluation criteria that are Root Mean Square Error (RMSE), and coefficient of determination (R^2). The proposed approach is depicted in **Figure 2**. The following subsections demonstrate the proposed approach.

3.1 L-RNN

A layered recurrent neural network is known as a neural network that has local feedback, which is particularly suited to predict the daily air pollutant attributes since it incorporates a time delay while training process through a feedback

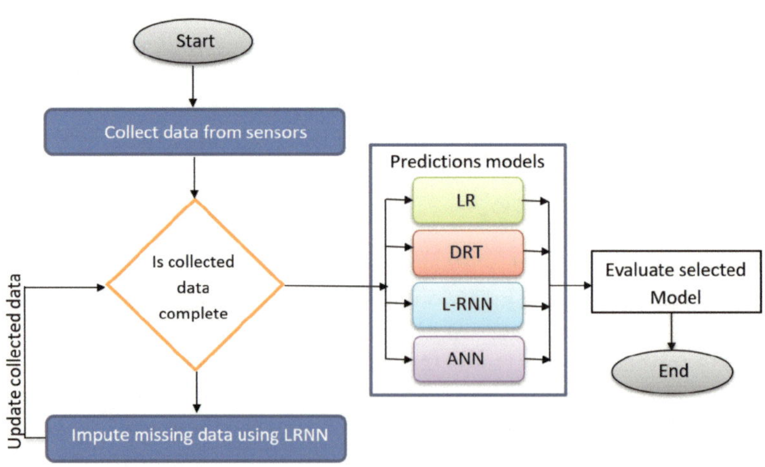

Figure 2.
The proposed approach.

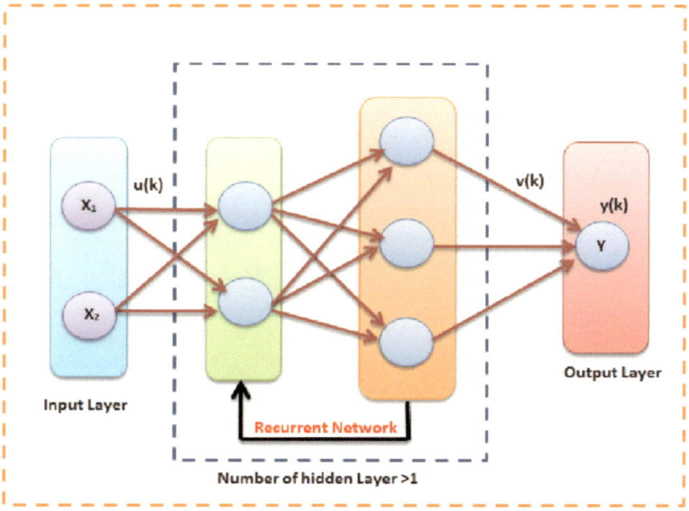

Figure 3.
Layer recurrent neural network (L-RNN).

connection between output layer and hidden layer(s). **Figure 3** demonstrates the connection feedback. In simple, during the training process, the output of the recurrent neural network is added to the output of the hidden layer. The result of summation is employed as an argument of the transfer function to gain the output in the succeeding iteration. Eq.(1) demonstrates the output of the L-RNN, where $u(k)$ presents the input values for hidden layer, $v(k)$ presents the input values for output layer. $W_{u,i}$ and $W_{v,j}$ presents the weights between u and v, respectively. The final output $y(k)$ is obtained from Eq.(2), where $f()$ is a transfer function. In this work, we employed back-propagation through time in the training phase for the proposed L-RNN structure.

$$v(k+1) = \sum_{i=0}^{n} W_{u,i}(k)u(k) + \sum_{j=0}^{m} w_{v,j}(k)v(k) \tag{1}$$

$$y(k) = f(v(k)) \quad where: f(v(k)) = \frac{1}{1 + exp\,(-(v(k))} \tag{2}$$

3.2 Data imputation using L-RNN

To implement the data imputation process, we clustered the data into two groups 1) complete dataset [without a missing value(s)] and 2) incomplete dataset [with a missing value(s)]. A holdout method is used to train and test the L-RNN. The complete dataset is divided into three datasets: training dataset (70%), testing dataset (15%), and validation dataset (15%). While the incomplete dataset is used to simulate the trained L-RNN model to impute the missing value(s). This process will be repeated dynamically while receiving any records with a missing value(s).

The computational complexity of the model depends on the structure of the L-RNN and the number of missing data in the received record. The computational complexity will increase exponentially if the number of missing values increases. **Figure 4** illustrates the process used to impute the missing value(s) (i.e., the concentration value of O_3, NO_2, $PM_{2.5}$, PM_{10}).

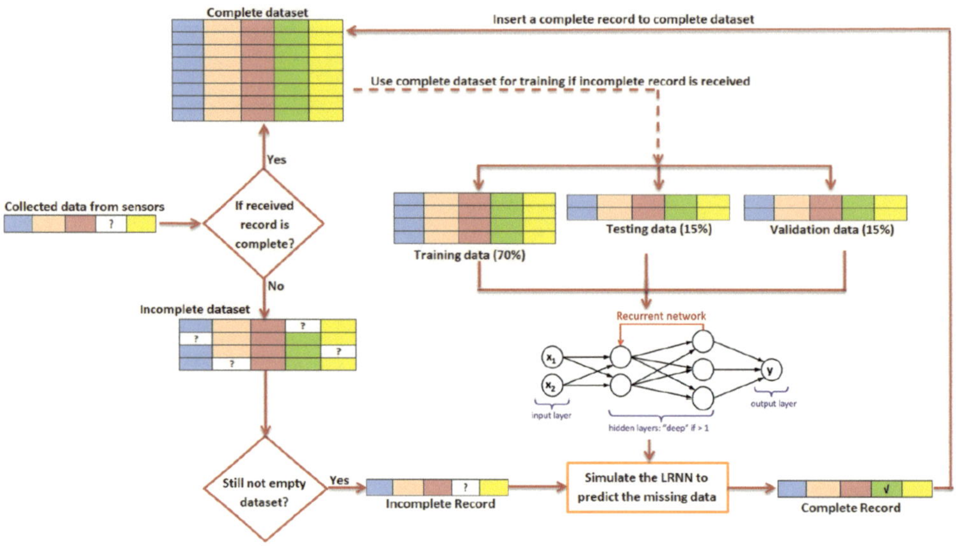

Figure 4.
Impute missing data approach.

4. Predictive models using machine learning

Several machine learning methods can be used to predict air quality. However, we have limited our research paper into four methods: MLR, DTR, ANN, and L-RNN. To avoid the over-fitting problem in the training process, we employed the k-fold cross-validation method with k-fold = 5. The following subsection explores each learning method in more detail.

4.1 Multiple linear regression

Linear regression (LR) is one of the most well-known algorithms in statistics and machine learning. The main idea of LR is to find the relationship between input and output numerical variables. There are several types of LR such as Simple linear regression, multiple linear regression, logistic regression, ordinal regression, Multinomial regression, and Discriminant Analysis. LR has been employed successfully in many areas as a machine learning algorithm [53, 54]. MLR is a classical statistical method that tries to find a relationship between complex input–output variables. In simple, MLR tries to find an approximation linear function between independent input variables and dependent output variable without loss of generality. Eq.(3) explores the regression line in MLR.

$$y = \beta_0 + \beta_1 x_1 + \dots + \beta_i x_i + \dots + \beta_k x_k + \varepsilon \quad (3)$$

where y is dependent output variable, x_i is the ith independent input variable, β_i is polynomial coefficients of x_i, k is the number of independent input variables, and $\varepsilon$is the possible variation form. Eq.(4) presents a compact version of Eq.(3).

$$y = X\beta + \varepsilon \quad (4)$$

where

$$y = \begin{bmatrix} y_1 \\ y_2 \\ \cdot \\ \cdot \\ \cdot \\ y_n \end{bmatrix}, \varepsilon = \begin{bmatrix} \varepsilon_1 \\ \varepsilon_2 \\ \cdot \\ \cdot \\ \cdot \\ \varepsilon_n \end{bmatrix}, \beta = \begin{bmatrix} \beta_1 \\ \beta_2 \\ \cdot \\ \cdot \\ \cdot \\ \beta_n \end{bmatrix}, and \quad X = \begin{bmatrix} 1, x_{1,1}, \dots, x_{1,k} \\ 1, x_{2,1}, \dots, x_{2,k} \\ \cdot \\ \cdot \\ \cdot \\ 1, x_{n,1}, \dots, x_{n,k} \end{bmatrix}_{n \times (k+1)} \tag{5}$$

where n represents the number of samples, $x_{m,i}$ represents the value of ith independent input variable in the mth sample, and ε_i is the ith residual error in the mth sample. The coefficient vector β can be calculated based on the standard least-square method as shown in Eq.(6).

$$\beta = \left(X^T X\right)^{-1} X^T Y \tag{6}$$

Therefore, when the parameter vector β is known, the generated MLR model can predict the dependent output variable based on the independent input variable (s).

4.2 Decision tree regression

The DTR is employed in this chapter to predict the air pollutant attributes due to its ability to handle complex data and takes less training execution time compared to other prediction models. In simple, DTR uses if-then conditions to predict the appropriate output value(s) [55]. The DTR has three steps to predict the output value(s) as follows:

- **Step 1**: Determining the parameter settings for DTR such as: predicting accuracy, selecting splits, when to stop splitting, and selecting the optimal tree.
- **Step 2**: Selecting the splits to predict values of the continuous dependent variable, which usually measured with node impurity measure which provides an indication of the relative homogeneity of cases in the terminal nodes.
- **Step 3**: Determining when to stop the splitting which is related to the minimum number of nodes. Which means to select the best rightly-sized tree, which is called the optimum tree.

4.3 Artificial neural network

Artificial Neural Network (ANN) has been used widely in many forecasting applications due to its ability to handle complex data. Without having any information about the mathematical model that represents the relation between input and output variables, ANN can learn the learn hidden knowledge between input and output variables. In general, there are many kinds of ANNs such as Feedforward Neural Network (FFNN), Recurrent Neural Network (RNN), and Convolutional Neural Network (CNN) [56]. In this chapter, we adopted two types of neural networks based on a feed-forward network using the propagation training method, which is: the standard neural network (ANN) and Layered Recurrent Neural Network (LRNN).

5. Data collection

The data set used in this research is collected from Dubrovnik city that is located in the east of Croatia. Dubrovnik city has a Mediterranean climate and has over 2600 hours of sunshine per year, which is considered the sunniest place in Croatia. In this dataset, the concentration of O_3 has been monitored with a commercial Teledyne API 400E UV photometric O_3 analyzer. While the concentration of NO_2 has been monitored with Teledyne API 200E chemiluminescent NO_2 analyzer. O_3 and NO_2 concentrations were measured every minute and the output signals were stored in a datalogger. The collected data are validated and averaged. The concentration of PM_{10}, and $PM_{2.5}$ have been monitored with the GRIMM model EDM 180. Samples of PM particles were collected by gravimetric methods throughout the day to obtain 24–hour averages of concentrations. All instruments are regularly maintained and calibrated. Meteorological data were obtained from the Meteorological and Hydrological Services of Croatia. The dataset is collected during the 2015 and 2016.

Table 1 shows the number of records in each dataset used in this paper. For example, the O_3 dataset has 699 total records, where 200 records (28.80%) are incomplete. **Figure 5b** demonstrates the missing data pattern for each dataset, where the x-axis presents the 24-hours (i.e., input variables), while the y-axis presents the observations during the 2 years. **Figure 5a** shows that there is a missing data in the second year for NO2 dataset, where NO_2 sensors do not work. Since the missing data are higher than 5%, we examined the collected data carefully to maintain the performance of air quality prediction systems. As a result, imputing missing data are needed.

6. Evaluation criteria

In this research, we employed two different evaluation criteria: Root Mean Square Error (RMSE), and coefficient of determination (R^2), defined below.

$$RMSE = \sqrt{\frac{1}{s}\sum_{i=1}^{s}\left(y_{i\,predicted} - y_{i\,observed}\right)^2} \tag{7}$$

$$R^2 = 1 - \frac{\sum_{i=1}^{s}\left(y_{i\,predicted} - y_{i\,observed}\right)^2}{\sum_{i=1}^{s}\left(y_{i\,observed} - \hat{y}_{i\,observed}\right)^2} \tag{8}$$

Dataset	$PM_{2.5}$	PM_{10}	O_3	NO_2
InComplete	179	179	200	270
Complete	551	551	499	461
Percentage of missing data %	16.96	17.08	20.26	28.80
Total number of records	730	731	699	731

Table 1.
Number of samples in each dataset.

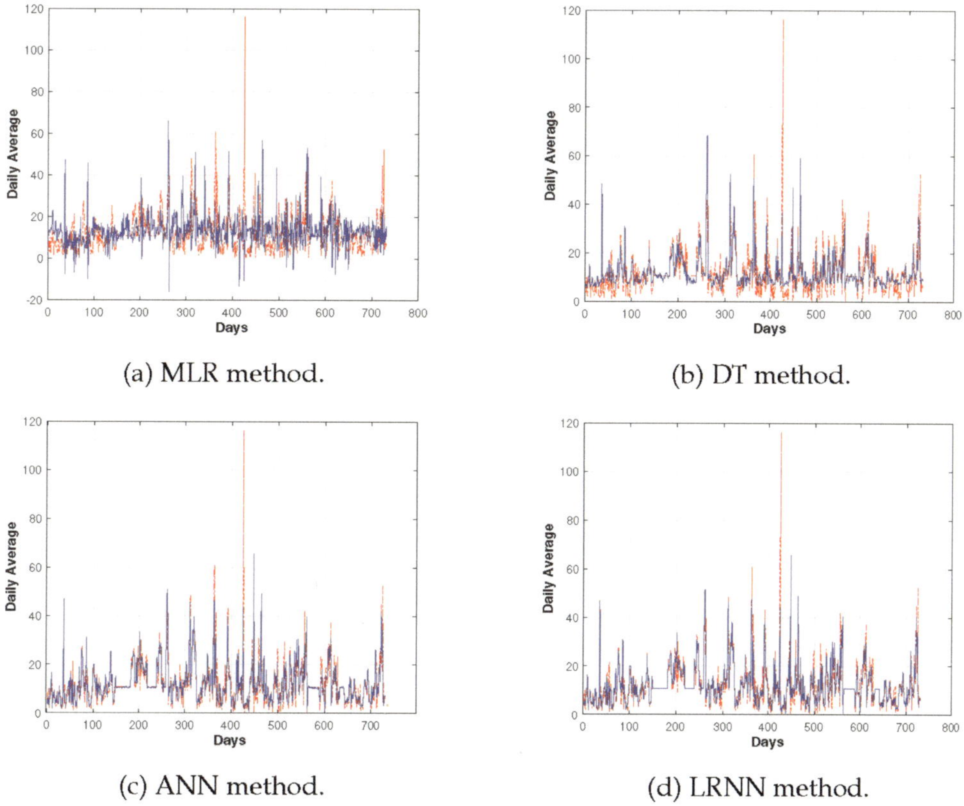

Figure 5.
Actual (–) and predicted (–.) values for No_2 *using all regression methods for* NO_2 *dataset.*

where $y_{i\,observed}$ and $y_{i\,predicted}$ denote the actual and predicted values of air pollution concentrations, respectively, s represents the number of instances and $\hat{y}_{i\,observed}$ stands for the average of the actual values of the air pollution concentrations.

Eqs.(7) and (8) show the evaluation process for each criteria. The minimum value of RMSE means better forecasting, while the maximum value of R^2 means better forecasting.

7. Experimental results

In this work, two different types of experiments were performed to develop a prediction model for pollutant parameters with missing data. They are: (i) removing missing or incomplete records, and (ii) imputing the missing data. Four regression models were employed in this work (i.e., MLR, DT, ANN, and LRNN). All experiments were performed using MATLAB-R2019b environment. The following subsections discussed the obtained results.

7.1 Results without imputing missing data

The first experiments that we employed in this chapter are based on removing all the missing data (i.e., records). **Table 2** shows the obtained results of four different regression models. The LRNN model outperforms other models in three

Regression model	NO_2		O_3		PM_{10}		$PM_{2.5}$	
	RMSE	R^2	RMSE	R^2	RMSE	R^2	RMSE	R^2
MLR	1.79	0.10	22.11	0.05	3.68	0.83	2.61	0.81
DT	1.73	0.16	20.44	0.19	3.68	0.83	2.57	0.82
LRNN	**0.26**	**0.85**	10.06	0.61	**0.30**	**0.88**	**2.39**	**0.90**
ANN	1.02	0.74	**8.39**	**0.76**	3.85	0.67	2.92	0.85

Table 2.
Results without imputing missing data.

datasets (i.e., NO_2, PM_{10}, and $PM_{2.5}$) based on RMSE and R^2 values. While ANN outperforms other models in O_3 dataset based on RMSE. The performance of the MLR method is the worst overall datasets.

7.2 Imputing data using LRNN

For imputing missing data, we employed L-RNN as a dynamic prediction model based on the current states of the collected data. In general, there are two different training algorithms for L-RNN: real-time recurrent learning, where a fixed set of weights recursively applied while training process and back-propagation through time, where the L-RNN structure altered between feed-forward and feedback structures. In this work, we used back-propagation through a time training process.

The parameters setting used, in this case, are shown in **Table 3**. A holdout method is employed to train the L-RNN based on the complete dataset, where 70% for training, 15% for validation, and 15% for testing. The reason for employing the holdout method is to reduce the complexity and execution time for the proposed imputing model. After imputing missing data, we employed a k-fold across-validation method in the training process for four machine learning methods (i.e., MLR, DTR, L-RNN, and ANN) with k-fold = 5 to evaluate the complete dataset.

7.3 Results after imputing missing data

7.3.1 MLR models

In this part, we employed MLR as a prediction model after imputing missing data. In Eq.(9), Eq.(10), Eq.(11), and Eq.(12) we show the MLR results for $PM_{2.5}$,

Parameters	Values
Number of iterations	1000
Number of neurons in the input layer	Number of input data
Number of neurons in the hidden layer	Number of input data /2
Number of neurons in the output layer	1

Table 3.
Parameter settings for the L-RNN model during imputing missing data.

DOI: http://dx.doi.org/10.5772/intechopen.93678

Dataset	MLR results	
	RMSE	R^2
NO_2	1.61	0.79
O_3	1.78	0.82
PM_{10}	3.84	0.46
$PM_{2.5}$	1.76	0.62

Table 4.
MLR results after imputing missing data.

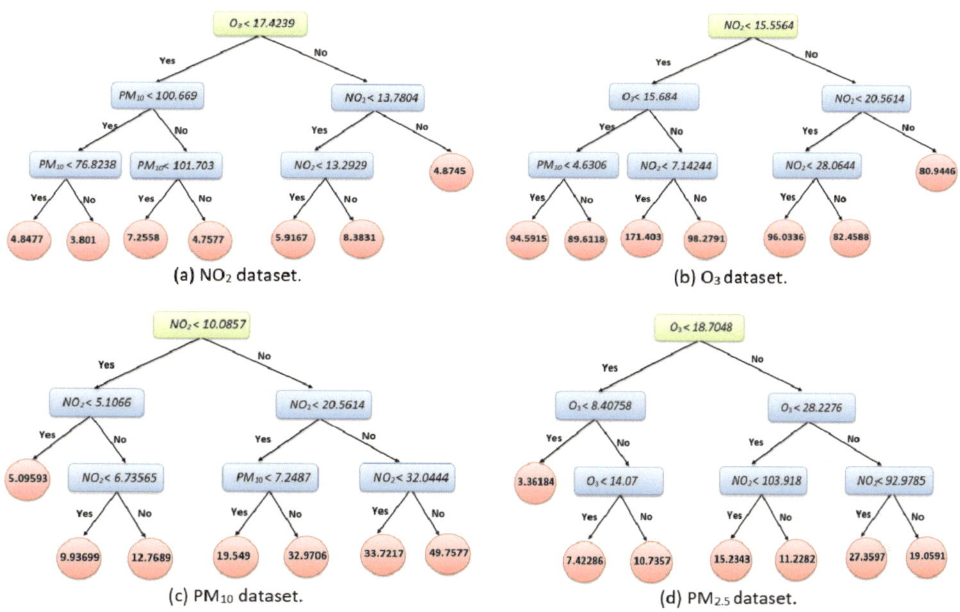

Figure 6.
Obtained tree for all datasets after imputing missing data.

PM_{10}, O_3, and NO_2, respectively. **Table 4** shows the obtained results of MLR method. The performance of of MLR is acceptable over all datasets.

$$PM_{2.5} = 4.4447 - 0.3753 \times NO_2 + 0.61551 \times PM_{10} - 0.025394 \times O_3 \tag{9}$$

$$PM_{10} = -3.8704 + 0.6183 \times NO_2 + 0.033796 \times O_3 + 1.2886 \times PM_{2.5} \tag{10}$$

$$O_3 = 95.039 - 0.22434 \times NO_2 + 0.96905 \times PM_{10} - 1.493 \times PM_{2.5} \tag{11}$$

$$NO_2 = 93.9595 + 0.0017985 \times O_3 + 0.15471 \times PM_{10} - 0.17977 \times PM_{2.5} \tag{12}$$

7.3.2 DT models

In this work, the minimum leave size used is 4, and the maximum number of splits is 6. The main reason for using this setting is to simplify the generated tree.

Dataset	DT results	
	RMSE	R^2
NO_2	2.36	0.65
O_3	7.32	0.54
PM_{10}	3.21	0.89
$PM_{2.5}$	3.14	0.85

Table 5.
DT results after imputing missing data.

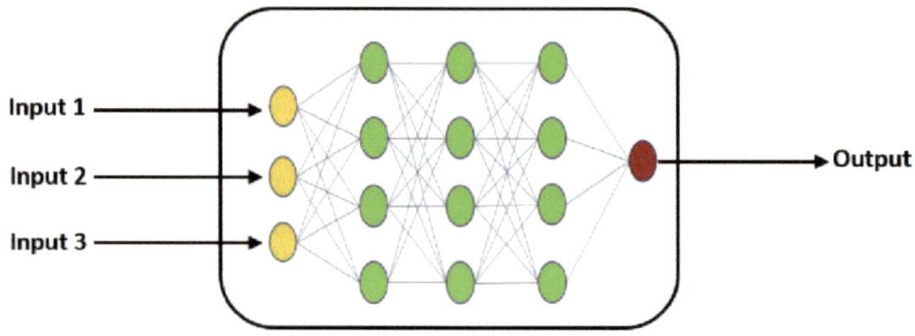

Figure 7.
ANN block diagram structure.

Dataset-	ANN results	
	RMSE	R^2
NO_2	0.05	0.96
O_3	3.25	0.78
PM_{10}	0.33	0.82
$PM_{2.5}$	0.16	0.97

Table 6.
ANN results after imputing missing data.

	Parameters	Value
LRNN	Number of epoch	1000
	Layer delays	1:2
	Hidden sizes	10
	Training function	Back propagation

Table 7.
Parameters setting for LRNN as a regression method.

The obtained models of DT for each dataset were shown in **Figure 6**. **Table 5** explores the obtained results of DT over all datasets.

7.3.3 ANN models

Figure 7 shows the ANN structure used in this chapter, where we have three inputs and a single output. **Table 6** shows the obtained results of

Dataset	LRNN results	
	RMSE	R^2
NO_2	0.22	0.93
O_3	2.76	0.80
PM_{10}	0.02	0.98
$PM_{2.5}$	0.02	0.93

Table 8.
LRNN results after imputing missing data.

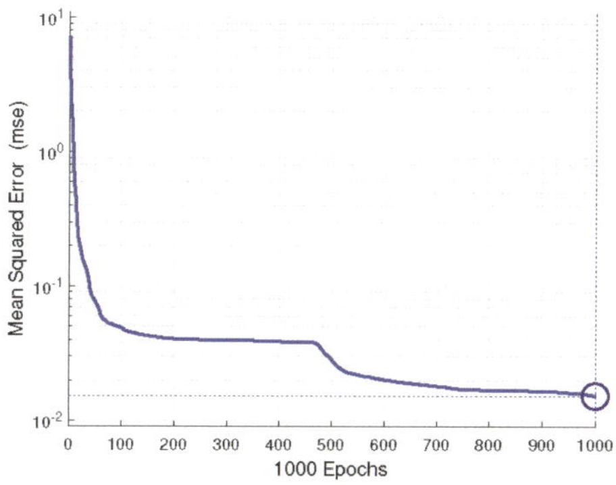

(a) NO_2 dataset.

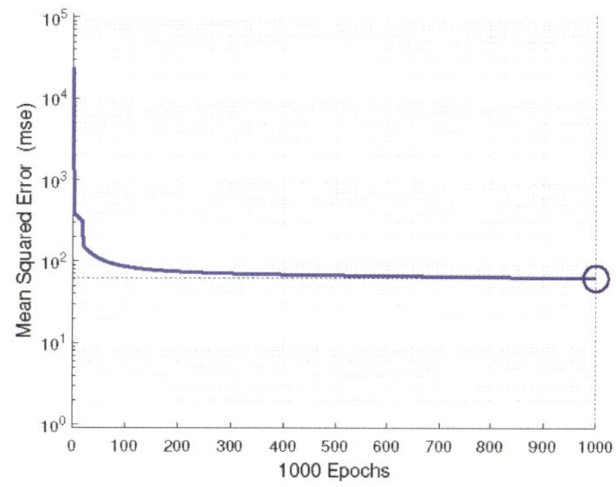

(b) O_3 dataset.

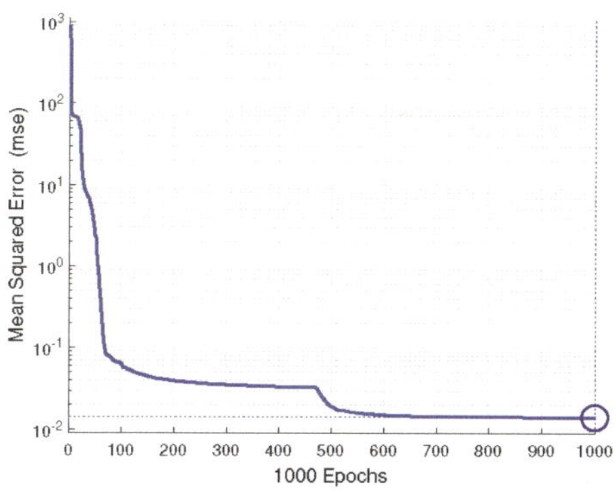

(c) PM_{10} dataset.

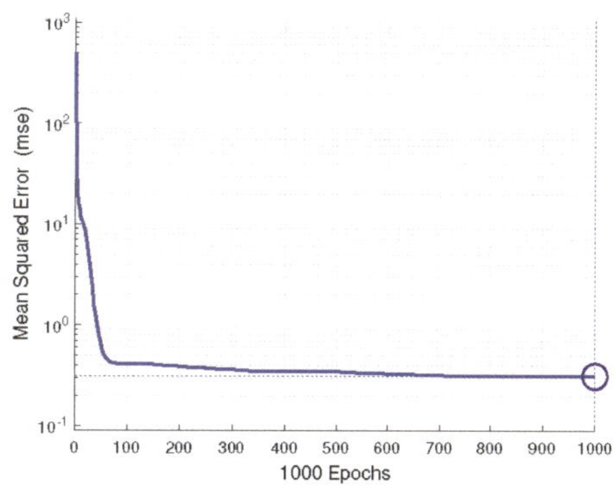

(d) $PM_{2.5}$ dataset.

Figure 8.
Convergence curves for LRNN over all datasets.

ANN over all datasets. The performance of ANN is excellent compared to MLR and DT.

7.3.4 LRNN models

In this chapter, we employed the LRNN as a regression model to predict the daily average of air pollutant attributes. **Table 7** shows the parameters setting for LRNN as a regression method. These settings have been selected carefully to fit our data based on a set ore preliminary experiments. **Table 8** shows the obtained results of LRNN. The performance of LRNN is outstanding based on the convergence curves as shown in **Figure 8**. LRNN method can converge within 1000 epochs.

Dataset	Regression model	After imputing		Without imputing	
		RMSE	R^2	RMSE	R2
NO_2	MLR	1.61	0.79	1.79	0.1
	DT	2.36	0.65	1.73	0.16
	LRNN	0.22	0.93	0.26	0.85
	ANN	**0.05**	**0.96**	1.02	0.74
O_3	MLR	1.78	0.82	22.11	0.05
	DT	7.32	0.54	20.44	0.19
	LRNN	**2.76**	**0.80**	10.06	0.61
	ANN	3.25	0.78	8.39	0.76
PM_{10}	MLR	3.84	0.46	3.68	0.83
	DT	3.21	0.89	3.68	0.83
	LRNN	**0.02**	**0.98**	0.3	0.88
	ANN	0.33	0.82	3.85	0.67
$PM_{2.5}$	MLR	1.76	0.62	2.61	0.81
	DT	3.14	0.85	2.57	0.82
	LRNN	**0.02**	**0.93**	2.39	0.9
	ANN	0.16	0.97	2.92	0.85

All significance values are in bold.

Table 9.
Results before and after imputing missing data.

Moreover, the obtained results of LRNN compared to the other previous methods are promising.

7.4 Analysis of the results

Table 9 shows the obtained results before and after imputing missing data. The performance of the LRNN model outperforms other models in three datasets (i.e., NO_2, PM_{10}, and $PM_{2.5}$) based on RMSE and R^2 values. While ANN outperforms other models in O_3 dataset based on RMSE. The performance of ANN over O_3 outperforms other methods. While the performance of MLR is the worst one.

From the obtained results, it can be seen that the performance of the LRNN model has an outstanding performance, where R^2 equals 0.90 in three datasets. However, these obtained results are not perfect since 16.96% of the data is removed for $PM_{2.5}$, and 28.80% of the data is removed from NO_2. Removing the missing data will neglect several records and the dataset may lose important information. **Figure 5** shows the actual and predicted values for NO_2 dataset using all regression methods after imputing missing data.

For more analysis, comparing the obtained results that are reported in **Table 9**, we can notice that the performance of MLR over PM_{10} after imputing the missing data is reduced 19%, while the performance of DT, LRNN, and ANN is improved after imputing missing data for PM_{10} dataset. In general, the performance of the regression models is improved compared to the results reported in **Table 2**.

For example, the R^2 value of ANN over O_3 dataset before imputing missing data was 0.76, and after imputing missing data becomes 0.78, while the RMSE is improved 39%. So, we can conclude that imputing missing data will improve the air quality measurement systems without losing any record of collected data.

8. Conclusion and future work

Data collection from remote sensors suffers from missing data which reduces the overall performance of air quality monitoring systems. Monitoring air pollution is not an easy task, where several measurements are used to evaluate air quality. In this study, four measurements are used to predict air pollution concentrations (i.e., O_3, NO_2, $PM_{2.5}$, and PM_{10}). We imputed the missing data using the Layered recurrent neural network (L-RNN). The performance of four different machine learning models (i.e., LR, DTR, ANN, and L-RNN) was investigated to predict the average daily air pollution concentrations. The performance of the proposed method presented an improvement in the performance of the air quality monitoring system. In future work, we plan to study different methods based on machine learning concepts to enhance the prediction of air pollutant systems. Moreover, we will investigate the general design of the Internet of Things (IoT) applications to improve the performance of the air quality monitoring system.

Acknowledgements

The authors would like to acknowledgement Croatian Meteorological and Hydrological Service for their support.

Author details

Hamza Turabieh[1*], Alaa Sheta[2], Malik Braik[3] and Elvira Kovač-Andrić[4]

1 Department of Information Technology, College of Computers and Information Technology, Taif University, Taif, Saudi Arabia

2 Computer Science Department, Southern Connecticut State University, New Haven, United States of America

3 Department of Computer Science, Al-Balqa Applied University, Salt, Jordan

4 Department of Chemistry, Josip Juraj Strossmayer University of Osijek, Osijek, Croatia

*Address all correspondence to: turabieh@gmail.com

References

[1] Delfino RJ, Staimer N, Tjoa T, Gillen D, Kleinman MT, Sioutas C, et al. Personal and ambient air pollution exposures and lung function decrements in children with asthma. Environmental Health Perspectives. 2008;**116**(4): 550-558

[2] Belwal C, Sandu A, Constantinescu EM. Adaptive resolution modeling of regional air quality. In: Proceedings of the 2004 ACM Symposium on Applied Computing, SAC '04. New York, NY, USA: ACM; 2004. pp. 235-239

[3] Dastoorpoor M, Goudarzi G, Khanjani N, Idani E, Aghababaeian H, Bahrampour A. Lag time structure of cardiovascular deaths attributed to ambient air pollutants in Ahvaz, Iran, 2008–2015. International Journal of Occupational Medicine and Environmental Health. 2018;**31**(4): 459-473

[4] Adhikari A. Chapter 1 - introduction to spatiotemporal variations of ambient air pollutants and related public health impacts. In: Li L, Zhou X, Tong W, editors. Spatiotemporal Analysis of Air Pollution and Its Application in Public Health. Netherlands: Elsevier; 2020. pp. 1-34

[5] Ghaly A. Mapping environmental pollution, contamination, and waste in the United States. In: Proceedings of the 3rd International Conference on Computing for Geospatial Research and Applications. United States: ACM; 2012. p. 41

[6] Chen Y, Wild O, Conibear L, Ran L, He J, Wang L, et al. Local characteristics of and exposure to fine particulate matter (pm2.5) in four Indian megacities. Atmospheric Environment: X. 2020;**5**:100052

[7] Gualtieri M, Øvrevik J, Holme JA, Perrone MG, Bolzacchini E, Schwarze PE, et al. Differences in cytotoxicity versus pro-inflammatory potency of different pm fractions in human epithelial lung cells. Toxicology In Vitro. 2010;**24**(1):29-39

[8] Milojevic A, Wilkinson P, Armstrong B, Bhaskaran K, Smeeth L, Hajat S. Short-term effects of air pollution on a range of cardiovascular events in England and wales: Case-crossover analysis of the minap database, hospital admissions and mortality. Heart. 2014; **100**(14):1093-1098

[9] Dastoorpoor M, Sekhavatpour Z, Masoumi K, Mohammadi MJ, Aghababaeian H, Khanjani N, et al. Air pollution and hospital admissions for cardiovascular diseases in Ahvaz, Iran. Science of the Total Environment. 2019; **652**:1318-1330

[10] Noel De Nevers. Air Pollution Control Engineering. Waveland Press. 2010

[11] Nowak DJ, Hirabayashi S, Doyle M, McGovern M, Pasher J. Air pollution removal by urban forests in canada and its effect on air quality and human health. Urban Forestry & Urban Greening. 2018;**29**:40-48. Wild urban ecosystems: challenges and opportunities for urban development

[12] Kovać-Andrić E, Sheta A, Faris H, Gajdosik MS. Forecasting ozone concentrations in the east of Croatia using nonparametric neural network models. Journal of Earth System Science. 2016;**125**(07)

[13] Sarwar G, Godowitch J, Henderson BH, Fahey K, Pouliot G, Hutzell WT, et al. A comparison of atmospheric composition using the carbon bond and regional atmospheric chemistry mechanisms. Atmospheric Chemistry and Physics. 2013;**13**(19):9695-9712

[14] Sheta A, Faris H, Rodan A, Kovač-Andrić E, Al-Zoubi A. Cycle reservoir with

regular jumps for forecasting ozone concentrations: Two real cases from the east of Croatia. Air Quality, Atmosphere and Health. 2018;**11**(03):559-569

[15] Fuks KB, Woodby B, Valacchi G. Skin damage by tropospheric ozone. Der Hautarzt. 2019:1-5

[16] Lange SS, Mulholland SE, Honeycutt ME. What are the net benefits of reducing the ozone standard to 65 ppb? An alternative analysis. International Journal of Environmental Research and Public Health. 2018;**15**(8)

[17] Isiugo K, Jandarov R, Cox J, Ryan P, Newman N, Grinshpun SA, et al. Indoor particulate matter and lung function in children. Science of the Total Environment. 2019;**663**:408-417

[18] Faustini A, Stafoggia M, Williams M, Davoli M, Forastiere F. The effect of short-term exposure to o3, no2, and their combined oxidative potential on mortality in Rome. Air Quality, Atmosphere and Health. 2019;**12**(5):561-571

[19] Kim C, Hu S-C. Total respiratory tract deposition of fine micrometer-sized particles in healthy adults: Empirical equations for sex and breathing pattern. Journal of Applied Physiology. 2006;**101**:401-412

[20] Deng Q, Lu C, Li Y, Sundell J, Norbäck D. Exposure to outdoor air pollution during trimesters of pregnancy and childhood asthma, allergic rhinitis, and eczema. Environmental Research. 2016;**150**:119-127

[21] Ul-Saufie A, Yahya A, Ramli N, Hamid H. Robust regression models for predicting PM10 concentration in an industrial area. International Journal of Engineering and Technology. 2012;**2**(3): 364-370

[22] Holgate ST, Koren HS, Samet JM, Maynard RL. Air Pollution and Health. United States: Elsevier; 1999

[23] Pokric B, Kreo S, Drajic D, Pokric M, Jokic I, Stojanovic MJ. Ekonet - environmental monitoring using low-cost sensors for detecting gases, particulate matter, and meteorological parameters. In: 2014 Eighth International Conference on Innovative Mobile and Internet Services in Ubiquitous Computing. United Kingdom: IMIS-2014, Conference Publishing Service (CPS); 2014. pp. 421-426

[24] Wang F, Liu J. Networked wireless sensor data collection: Issues, challenges, and approaches. IEEE Communication Surveys and Tutorials. 2011;**13**(4):673-687

[25] Turabieh H, Abu Salem A, Abu-El-Rub N. Dynamic L-RNN recovery of missing data in iomt applications. Future Generation Computer Systems. 2018;**89**:575-583

[26] Yu Y, Si X, Hu C, Zhang J. A review of recurrent neural networks: LSTM cells and network architectures. Neural Computation. 2019;**31**(7):1235-1270

[27] Choi E, Schuetz A, Stewart W, Sun J. Using recurrent neural network models for early detection of heart failure onset. Journal of the American Medical Informatics Association. 2016; **24**:ocw112

[28] Oeda S, Kurimoto I, Ichimura T. Time series data classification using recurrent neural network with ensemble learning. In: Gabrys B, Howlett RJ, Jain LC, editors. Knowledge-Based Intelligent Information and Engineering Systems. Berlin Heidelberg: Springer; 2006

[29] Che Z, Purushotham S, Cho K, Sontag D, Liu Y. Recurrent neural networks for multivariate time series with missing values. Scientific Reports. 2016;**8**:06

[30] Momeni A, Pincus M, Libien J. Imputation and Missing Data. United

States: Springer International Publishing; 2018. pp. 185-200

[31] Lang KM, Little TD. Principled missing data treatments. Prevention Science. 2018;**19**(3):284-294

[32] Mary IPS, Arockiam L. Imputing the missing data in iot based on the spatial and temporal correlation. In: 2017 IEEE International Conference on Current Trends in Advanced Computing (ICCTAC). Netherlands: Elsevier; 2017. pp. 1-4

[33] Sta HB. Quality and the efficiency of data in "smart-cities". Future Generation Computer Systems. 2017;**74**: 409-416

[34] Feng X, Wu S, Liu Y. Imputing missing values for mixed numeric and categorical attributes based on incomplete data hierarchical clustering. In: Xiong H, Lee WB, editors. Knowledge Science, Engineering and Management. Berlin, Heidelberg: Springer Berlin Heidelberg; 2011. pp. 414-424

[35] Sen S, Das M, Chatterjee R. Estimation of incomplete data in mixed dataset. In: Sa PK, Sahoo MN, Murugappan M, Wu Y, Majhi B, editors. Progress in Intelligent Computing Techniques: Theory, Practice, and Applications. Singapore: Springer Singapore; 2018. pp. 483-492

[36] Chen M, Hao Y, Hwang K, Wang L, Wang L. Disease prediction by machine learning over big data from healthcare communities. IEEE Access. 2017;**5**: 8869-8879

[37] Perez P, Gramsch E. Forecasting hourly pm2.5 in santiago de chile with emphasis on night episodes. Atmospheric Environment. 2016;**124**: 22-27

[38] Laña I, Del Ser J, Padró A, Vélez M, Casanova-Mateo C. The role of local urban traffic and meteorological conditions in air pollution: A data-based case study in Madrid, Spain. Atmospheric Environment. 2016;**145**: 424-438

[39] Kamińska JA. The use of random forests in modeling short-term air pollution effects based on traffic and meteorological conditions: A case study in wrocław. Journal of Environmental Management. 2018;**217**:164-174

[40] Kamińska JA. Probabilistic forecasting of nitrogen dioxide concentrations at an urban road intersection. Sustainability. 2018;**10**: 4213

[41] Shang Z, Deng T, He J, Duan X. A novel model for hourly pm2.5 concentration prediction based on cart and eelm. Science of the Total Environment. 2019;**651**:3043-3052

[42] Braik M, Sheta A, Al-Hiary H. Hybrid neural network models for forecasting ozone and particulate matter concentrations in the Republic of China. 13. Air, Quality, Atmosphere, and Health. 2020;**13**:839-851. Springer

[43] Sheta AF, Ghatasheh N, Faris H. 2015 6th International Conference on Information and Communication Systems (ICICS). Forecasting global carbon dioxide emission using auto-regressive with eXogenous input and evolutionary product unit neural network models. 2015;182-187. DOI: 10.1109/IACS.2015.7103224

[44] Dotse S-Q, Petra MI, Dagar L, De Silva LC. Application of computational intelligence techniques to forecast daily pm10 exceedances in Brunei Darussalam. Atmospheric Pollution Research. 2018;**9**(2):358-368

[45] Sun W, Sun J. Daily pm2.5 concentration prediction based on principal component analysis and lssvm optimized by cuckoo search algorithm.

Journal of Environmental Management. 2017;**188**:144-152

[46] Xu Y, Du P, Wang J. Research and application of a hybrid model based on dynamic fuzzy synthetic evaluation for establishing air quality forecasting and early warning system: A case study in China. Environmental Pollution. 2017; **223**:435-448

[47] Luo H, Wang D, Yue C, Liu Y, Guo H. Research and application of a novel hybrid decomposition-ensemble learning paradigm with error correction for daily pm10 forecasting. Atmospheric Research. 2018;**201**:34-45

[48] Aznarte JL. Probabilistic forecasting for extreme no2 pollution episodes. Environmental Pollution. 2017;**229**: 321-328

[49] Wang D, Wei S, Luo H, Yue C, Grunder O. A novel hybrid model for air quality index forecasting based on two-phase decomposition technique and modified extreme learning machine. Science of the Total Environment. 2017; **580**:719-733

[50] Kumar A, Goyal P. Forecasting of air quality in Delhi using principal component regression technique. Atmospheric Pollution Research. 2011;**2** (4):436-444

[51] Akhtar A, Masood S, Gupta C, Masood A. Prediction and analysis of pollution levels in Delhi using multilayer perceptron. In: Satapathy SC, Bhateja V, Raju KS, Janakiramaiah B, editors. Data Engineering and Intelligent Computing. Singapore: Springer Singapore; 2018. pp. 563-572

[52] Yadav V, Nath S. Identification of relevant stochastic input variables for prediction of daily pm10 using artificial neural networks. In: Ray K, Sharma TK, Rawat S, Saini RK, Bandyopadhyay A, editors. Soft Computing: Theories and Applications. Singapore: Springer Singapore; 2019. pp. 23-31

[53] Singh P. Linear Regression. Berkeley, CA: Apress; 2019. pp. 43-64

[54] Wang S, Huang GH, He L. Development of a clusterwise-linear-regression-based forecasting system for characterizing dnapl dissolution behaviors in porous media. Science of the Total Environment. 2012;**433**:141-150

[55] Swetapadma A, Yadav A. A novel decision tree regression-based fault distance estimation scheme for transmission lines. IEEE Transactions on Power Delivery. 2017;**32**(1):234-245

[56] Qin H, Gong R, Liu X, Bai X, Song J, Sebe N. Binary neural networks: A survey. Pattern Recognition. 2020;**105**:107281

Chapter 4

Wind Power Forecasting

Sumit Saroha, Sanjeev Kumar Aggarwal and Preeti Rana

Abstract

The wind power generation depends on wind speed and its derivatives like: wind speed and direction. With consideration of stochastic nature of wind power, this work addresses three main issues: first, it discusses the state of art of energy forecasting with emphasis on wind power forecasting. It provides an overview of different variables on which wind power generation depends and explains various key features regarding the design framework of forecasting models. Second, it performs an assessment, detailed comparison and evaluation of the forecasting performance of various types of models; and third, evaluates the uncertainty of expected outcomes with the help of probabilistic measures.

Keywords: forecasting, neural networks, probability, time series, wind power

1. Introduction

Electricity sector especially in supply industry over the last various years across the world has underwent through numerous structural and systematic changes due to two main reasons: orientation of industry towards privatizations (reforms) and movement of electricity generation towards clean and pollution free renewable energy sources [1]. In this changing environment forecasting electricity becomes one of the most important exercises in managing the power systems. Forecasting plays a significant role in operation planning, scheduling and real time balancing of power system. Mainly, there are three forecasting issues in present day power systems namely electricity load, price and the renewable energy sources. Among the recently emerged renewable sources of energy (solar energy), the wind power industry has witnessed tremendous growth and has taken a leading role [2, 3].

Besides this, the electricity based on renewable energy sources perceived as an alternate source of energy and their penetration within the power system is rising at a very fast rate [4]. Among new sources of renewable energy, the wind energy has seen tremendous growth over recent years; in various countries, it is a true alternative to fossil fuels. Furthermore, wind power generation capacity varies constantly, stochastic, intermittent in nature and associated with generation of other ramp events. In spite of that, it is freely available & pollution free source of energy; so, it has gained an extensive interest and one of the most established renewable energy alternatives to the conventional energy resources. On approaching towards the end of 2016, 486.8 GW would be worldwide installed wind nameplate capacity due to growth rate of 12.5%. As per estimate, wind power towards the end of 2021 will approach to 817 GW with growth rate of 10.4%. These wind capacity installations are mainly utilized in electric power systems based on large grid and their interconnections [5, 6]. Now-a-days another fast growing eco-friendly electrical generation technologies are solar, geothermal and tidal energy.

The uncertainty associated with wind power originates from uncertainties in its derivatives such as: wind speed & direction forecasts. In coordination with fast deployment of wind farms establishes a demand for efficient forecasting methods related to wind power production. The high is forecast reliability, low will be reserve maintenance cost of the system, which will result technical and commercial implications for proper management and working of power systems. Wind power forecasting (WPF) depicts how much wind power is to be expected at particular instant of time in the days to come. WPF is one of the most critical aspects in wind power integration and operation [6–8]. As per time horizons, the WPF has been done on the basis of long, medium and short term.

The availability of wind power is largely influenced by the prevailing weather conditions, seasonal variations and time spam variation and therefore, it is characterized by strong fluctuations, uncertainty and intermittency. These characteristics of wind power create a great attention towards it. Consequently, power generation from wind cannot be matched easily to the electricity demand like power generated with conventional plants. The penetration (share of wind power to meet demand) level of wind power introduces new challenges for the power system, some of them include:

Integration with Grid: The management of intermittence of wind generation is the key issue related to its integration with grid. The transmission utility is only responsible for the balancing of demand and supply at grid level. Therefore, it is necessary to schedule the supply in advance in order to meet the load profile. The load is corresponding to the total demand of electricity consumption over a definite area. The load forecast is usually given by the load forecasting models. The Mean Absolute Percentage Error (MAPE) of load is in the order of 0.87–1.34% [9] for the day ahead or week ahead predictions. Still continuous effort has been made by various researchers and practitioners for improving the performance of load forecasting models and techniques. i.e. it is reached in advance stage of research.

Integration with Electricity Markets: Generally, the electricity market is build by two mechanisms. The first one is spot energy market or so called Day Ahead market, where the bulk energy necessary to cover the load profile for the next coming day is traded on the generation cost. An auction process followed by bidding permits the settlement of electricity price and generation for the various bidding hours. The second mechanism is ancillary service market or so called intraday market, where differences between planned production and actual load are traded (due to the power plant failure or due to intermittence of wind power generation). The ancillary service market is very important for a stable operation of the power grid and span across various time frames. Therefore, it is additionally important for consumers as well as suppliers to know the future electricity price, so that they can make strategies. Like load forecasting the electricity price is in its advance stage of research and error rate (MAPE) reported is 3.96–4.92% [10].

Therefore, the accurate forecasts of wind power generation is an essential factor for a successful integration of large amounts of wind power into the electricity supply system, aiming at precise information on timing and magnitude of power generation from these variable sources.

Among requirements of wind power forecasting over three different forecasting horizons, there are different framework for the forecasting which includes single step ahead, multiple lead hours ahead and probabilistic forecasting. Typically multiple step and probabilistic forecasting is more complicated because in multiple, the error is multiples at every lead hour prediction; whereas, in probabilistic several statistical factors contribute additional complexity and additional complicacy. Moreover, it also affects the profits of a utility directly.

2. Methods of energy forecasting

2.1 Deterministic or point forecasting

The predicted values can be provided to end-users either in a deterministic or in probabilistic format, with the former, a specific value for energy production at a particular time step (15-minutes or one hour) is forecasted; whereas, in later, range of possible output is forecasted on the behalf of deterministic forecasted values using probability theory.

Single Step Ahead Forecasting.

It is the estimation of any quantity today for the next coming day with utmost possible precision and reliability. We have at our disposal the past values of this quantity, the data of one or several time series along with other several factors on which these time series are produced.

$$WP_{t+1} = f(WP_t, \dots \dots, WP_{t-d+1}) + e \quad (1)$$

With

$$t \in f\{d, \dots \dots, N-1\} \quad (2)$$

By the Eq. (1), e, is the prediction error or noise present between present forecasting value and n previous observations. WP is the wind power, T is the target, for multiple step the target matrix is increased with respect to each step in advance as given below in Eq. (3, 4).

$$\textit{Single Step} \begin{bmatrix} WP_{11} WP_{12} --- WP_{16} \\ WP_{21} WP_{22} --- WP_{26} \\ \\ WP_{N1} WP_{N2} --- WP_{N6} \end{bmatrix} \begin{bmatrix} T_1 \\ T_2 \\ \\ T_N \end{bmatrix} \quad (3)$$

$$\textit{Second Step} \begin{bmatrix} WP_{11} WP_{12} --- WP_{16} \\ WP_{21} WP_{22} --- WP_{26} \\ \\ WP_{N1} WP_{N2} --- WP_{N6} \end{bmatrix} \begin{bmatrix} T_2 \\ T_3 \\ \\ T_{N+1} \end{bmatrix} \quad (4)$$

Multi Step Ahead Forecasting.

The multiple steps ahead or multiple lead hour prediction is forecasting a pattern of values for given time series. It is an approach that works step-by-step by using current prediction for deterministic next stage prediction. In case of multi-step ahead prediction various anomalies like error accumulation and complexity of data prevails when prediction period is long. It all occurs due to propagation of bias and variances form previous prediction of future prediction. Because of this large forecasting horizon & error present in forecasting this method is suffered from the low performance & higher inaccuracy that is because of use of approximated values rather than actual values. The main reason for this higher inaccuracy is that the error is multiplied in every step-ahead prediction. So, the selection of input parameter function to fit the time series can be a challenging task for the power system researchers.

$$K^{th}\textit{Step} \begin{bmatrix} WP_{11} WP_{12} --- WP_{16} \\ WP_{21} WP_{22} --- WP_{26} \\ \\ WP_{N1} WP_{N2} --- WP_{N6} \end{bmatrix} \begin{bmatrix} T_K \\ T_{K+1} \\ \\ T_{N+K} \end{bmatrix} \quad (5)$$

2.2 Probabilistic or interval forecasting

The probabilistic forecast systems are designed to estimate the uncertainty of a forecast and used to produce the application of probabilistic forecasting. The verification is an essential part of probabilistic forecast systems. The correct and accurate use of probability forecasts means that, given a large sample, on average and event will occur at the same frequency as the forecast probability [11].

3. State of art for wind power forecast

As far as literature is concerned, number of forecasting methods have been designed and analyzed over last few decades. Based on information in research papers, author has examined various developments in the field of wind power generation & its derivatives prediction such as speed or direction. The major emphasis is led on facilitation of a number of issues concerned with techniques involved in WPF, focuses on complexity reduction in forecasting issues with higher accuracy in forecasting for different time span. This research mainly focuses on motivating power system researchers to design highly efficient and accurate models whether online/offline considering various issues related to wind power which in twin result in reliable operation of power system models by utilizing energy resources economically. On carrying out comparative study and analysis of accuracy in forecasting models, hybrid models outperformed all other models.

The generation of wind power is highly influenced by nature and seasons. So, it has been a tedious task to design a sound prediction model by taking in account above two factors. But, AI and machine learning have come with an advantage for developing new models due to their higher efficiency and accuracy. After a deep insight of various research papers authors have observed that the NN is the most prevailing approach for wind power and its derivatives estimation. It has also been observed that, hybrid models have been found to be more accurate model and for getting more accuracy, the training data should be updated regularly with small time span. Although for real time operation of power system, researchers have to move towards online models. There are three main steps involved in WPF (i) Input Selection, (ii) Data Pre-processing, & (iii) Forecasting models (tool) used.

3.1 Input parameters & their selection methods

The higher uncertainty in wind nature is result of uncertainties in its derivatives that affect systems of reliability. If forecast reliability is higher than operational cost of wind power system is lowered, in turn benefitting wind farm owners as they will have more substantial saving as well as have better efficiency of the system [12]. Apart from all this, wind power prediction is still a tedious task because wind flow is an unpredictable natural phenomenon and wind speed time series possesses various characteristics like: high volatility, high complexity, non linearity and non-stationary due to prevent physical conditions of place [13, 14]. After an extensive study of various research papers more than 46 exogenous variables have been observed as given in **Table 1**.

The input variables selection is main task because the accurate prediction by a forecasting model is highly influenced by proper input variables and their past results in the field of wind speed & power prediction and estimation. Furthermore, the selection of input variables for a prediction model mainly depends on exogenous and without exogenous variables. The various input selection techniques are as discussed.

Class	Input variable	Input data
1. Atmospheric Characteristics	(1) Temperature (2) Pressure, (3) Humidity (4) Rainfall, (5) Cloud formation, (6) Cloud cover, (7) Turbulance, (8) Radiations Effect, (9) Density	
2. Topographic Characteristics	(10) Turbine position, (11) Turbine size, (12) Hub height, (13) Tower height, (14) Elevation, (15) Degree in Latitude	
3. Wind Power Characteristics	(16) Wind speed, (17) Wind direction, (18) Radiation transmission, (19) Sine & Cosine of wind direction, (20) Air density, (21) Local wind profile	*f*(wind Speed); (d-m, t), m = 1,2,3,4,7,8, 168, 365
4. Behavior Indices	(22) Hydrological cycle, (23) cloud-radiation interaction, (24) spatial behavior, (25) Temporal behavior, (26) Spatial resolution	*f*(wind power; (d-m,t-n), m = 1,2,3,4,7,8, 168, 365 and n = 0,1,2,3,4
5. Other Stochastic Uncertainty	(27) Ocean-land interactions, (28) Regime switching, (29) Exchanges of momentum, (30) Load distribution among parallel turbines, (3) 1Thunders, (32) Storms, (33) Risk index, (34) Guest wind speed	*f*(wind direction; (d-m,t-n), m = 1,2,3, 168, 365 and n = 0,1,2,3,4
6. Geographical Conditions	(35) Orography, (36) Surface roughness, (37) Obstacles, (38) Geographical height, (39) Mean sea level pressure, (40) Air temperature, (41) Soil wetness, (42) Atmosphere covering, (43) Snow covering, (44) Moisture with land surface, (45) Complex terrain, (46) Terrain roughness	

Table 1.
Factors affecting wind power generation.

3.1.1 Physical or numerical weather prediction (NWP) models

These are very common model in which wind is a function of exogenous variables and forecasting tool input is the output of NWP models. These physical models forecasting process depends on entire input corresponding to wind power derivatives and are deterministic one. Their implementation process is very complex to perform, take high computation time to carry out forecasting process and depends on physical variables concerned with wind farm location. The equation which is used to convert wind speed into power is as follows as: $W_p = 0.5.\rho.A.v^3$. Here, ρ denotes the air density; v denotes the wind velocity through an intercepting area A of wind turbine. Actually, this equation follows the different physical variables corresponding to wind turbine. The purpose of NWP models is to predict the wind speed of surrounding area of wind mill.

3.1.2 Statistical models

In statistical models, wind remains a function that works using past captured values. These models are trained by providing data patterns that are measured statistically. They are based on historical data patterns generated by wind power and hence, they are not based on computation of any form of mathematical expression. These models outperform other short term forecasting horizon over prediction accuracy and these models are easy to implement & validate. They employed the statistics like: Cross Correlation (CC), Auto Correlation Function (ACF) and Partial Auto Correlation Function (PACF) for input selection on the basis of standard deviation, variance, mean and slope of input curve etc. The **Figure 1** shows ACF and PACF of hourly Wind Power time series based on these two parameters input

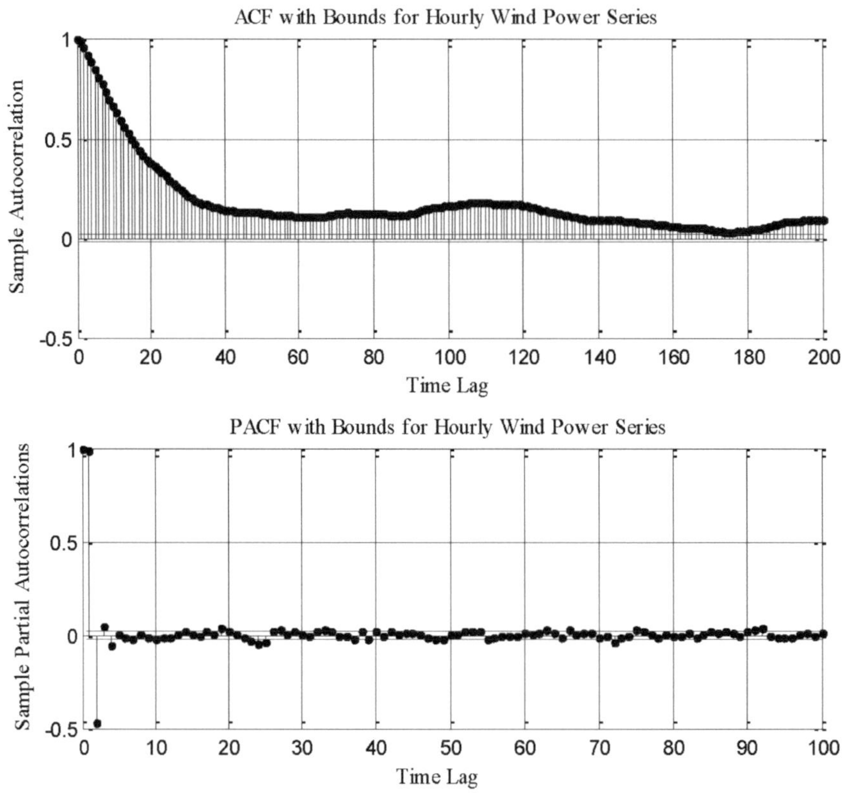

Figure 1.
ACF & PACF for hourly wind power series.

time lag parameterization of both time series and Artificial Intelligence (AI) take place. The higher is the value of ACF more is correlation between two consecutive series. However, the selection of input variables is one of the most important part of NN based forecasting model on with the accuracy of the model depends and that also determines the input architecture of the model. During the training of NN model, there may be problem of overtraining or over fitting that leads to poor accuracy of model. Therefore, it is necessary to know the relation that exists between present time wind power series along with their past time lag series. The input time lag is given below in **Table 2**. The wind forecast problem aims to find an estimate $WP(t + k)$ of the wind vector $WP(t + n)$ based on the previous n measurements $WP(t)$, $WP(t\text{-}1)$,…, $WP(t\text{-}n)$.

3.1.3 Hybrid (physical + statistical) models

It is the combination of NWP and statistical tools for input data selection. In this, on the bases of statistical analysis, the NWP variables are pre-processed to time lag for the prediction of next step.

3.2 Input data collection & pre-processing

The input data and wind data pattern is accumulated in raw form and does not possesses highly efficient forecasting capability with accurate precision. Raw data is unpredictable, irregular, seasonal and more complex due to changing weather. While prediction computation, over-fitting or over-training of NN is the main issue in time series variation leading to foot fall in accuracy of forecasted values. Data

S. No.	Time lag series	No. of time lag
1.	*WP (t-1)*	1
2.	*WP (t-1), WP (t-2)*	2
3.	*WP (t-1), WP (t-2), WP (t-3)*	3
4.	*WP (t-1), WP (t-2), WP (t-3), WP (t-4)*	4
5.	*WP (t-1), WP (t-2), WP (t-3), WP (t-4), WP (t-5)*	5
6.	*WP (t-1), WP (t-2), WP (t-3), WP (t-4), WP (t-5), WP (t-6)*	6
7.	*A1*	Approximate Series
8.	*D1, D2, D3, D4, D5, D6*	Detailed Series

Table 2.
Inputs used.

pre-processing means data cleaning data transformation and data reduction input data and converting it into useful information as per dimensions. Data must be classified based on seasonal and weather variable variation. Kalman filter is an appropriate solution to various problems such as: complexity in data, over-fitting and outliers of input data generated during learning process [15, 16]. As Unscented Kalman Filter (UKF) achieves higher efficiency in handling random fluctuations, so it is an economical and adequate choice for non-linear estimation of wind speed [17].

In presented work, in order to investigate the performance of different forecasting models, real wind generation data of Ontario Electricity Market (OEM) from 2011 to 2014 [18] has been considered. For obtaining more accuracy and over-training avoidance in learning process to achieve greater accuracy, large set of data values have not been considered, as generation of wind power is dependent function on numerous parameters such as: changing season, temperature and weather conditions. As time moves wind capacity (defined as actual energy produced in comparison to energy actually dissipated by turbines under favorable conditions) can fluctuate. The main concern of Wavelet Transform (WT) is to collect the meaningful information with removal of noise & irregularities from the original signal. From the available literature on forecasting and experimental analysis, it has been observed that Daubechies wavelet at different levels performs an appropriate smoothness of the signal with respect to wave-length, which results in an appropriate behavior of input data pattern for wind power prediction tool.

The WT implementation is done to decompose wind power series broadly into constitutive series set. This set of constitutive series help in reduction of input data and outperforms original wind series in behavior leading to prediction accuracy improvement. The WT divides wind series signal into two distinguishing signals having low and high frequency, then the decomposed signals are provided to the separate NN model for training. There are four filters (decomposition low pass & high pass filter, reconstruction low & high pass filter) used in Discrete Wavelet Transform (DWT) for scaling the input data pattern into approximate (A) and detailed (D) signals as given in **Table 2** [19–24]. Empirical Model Decomposition (EMD) has also been used to decompose the wind power series into high and low frequency signals [25]. The NN models train themselves better with the pre-processed data, as a result of this better prediction performance.

3.3 Wind power forecasting tools

For the past two decades, models based on machine learning have captured attention & become more sophisticated and reliable contenders in spite of traditional statistical models in forecasting. These are non parametric & non-linear models also known as data driven or black box models having usage of historical data patterns to learn the stochastic dependency between past and future. These NN's models always leave behind other traditional statistical models such as: linear regression and Box-Jenkins approaches. The NNs can be successfully used for modeling and forecasting non-linear time series [26].

3.3.1 Statistical models

The conventional statistical models (persistence, Moving Average & Gray Models) are identical to the direct random time-series model. Based on a number of historical data, pattern identification, parameter estimation, model checking are utilized to make a mathematical model for the prediction problem.

i. Traditional Models

a. **Naïve Predictor:** In order to get a significant evaluation of WPF a naïve model should be used. This is one of the old and simple ways to forecast wind power & speed also called persistence model. It is based on the simple assumption that wind power at present time t will be same in a future time (t + x) [27].

b. **Simple Moving Average:** The moving average predicts the wind power based on simply the average of past values of wind power. It has also been used as a benchmark for assessing the accuracy criteria of prediction model.

c. **Gray Model (1,1) Predictor:** GM (n, m) model is based on the Gray theory as demonstrated by Professor Deng in 1982. GM (n, m) denotes a Gray model where n is the order differential equation and m is the no. of variables. It predicts the future values of time series based on the recent data fluctuations. There are various types of Gray Models as designed by various researchers but because of computational efficiency of GM (1, 1) is generally used.

ii. Linear or Time Series (TS) Models

According to the methods which have been proposed by Jenkins, these models can be further divided as follows: autoregressive model (AR), moving average model (MA), autoregressive moving average model (ARMA), auto regressive integrated moving average model (ARIMA) [28]. Generalized Autoregressive Conditional Heteroskedasticity (GARCH) has been used for interval forecasting to simulate the fluctuating characteristics of the residual series in Mictrogrid China. Fractional-ARIMA method has been proposed to overcome the disadvantage of ARIMA method, which has been characterized by a slow decay in its ACF [29]. The stochastic and seasonality pattern of wind power has been tackled by designing a combined Autoregressive Fractionally Integrated Moving Average (ARFIMA) and GARCH model [30]; whereas, for above said problem ref. [31] demonstrated ARMA with Vector Auto-regression and ref. [32] designed different ARMA models for wind speed and direction tuples prediction (above said problem).

3.3.2 Artificial intelligence (AI) models

The FFNN architecture, which is also called as Multi Layer Perceptron (MLP), along with back propagation (BP) as the learning algorithm is the most popular choice among researchers. The neural network (NN) and machine learning algorithms structures used by most of the researchers after 2000 in the leading journals are: Feed Forward Neural Network (FFNN), Recurrent Neural Network (RNN), Radial Basis Function Neural Network (RBFNN), Support Vector Machine (SVM), Support Vector Regression (SVR), Adaptive Neuro Fuzzy Inference System (ANFIS), Extreme Learning Machine (ELM), Adaptive Wavelet Neural Network (AWNN), General Regression Neural Network (GRNN), and Linear Neural Network with Time Delay (LNNTD).

In this, wind forecasting has been done by the three different models: (i) Benchmark, (ii) NN and (iii) WT based model. In the first category, only Naïve Predictor has been considered. This is the standard benchmark for wind forecasting applications, in which the previous values of input wind power series have been used for the next lead hour as forecasted values. In the second category, different ANN based models have been taken into consideration with different structure of network and learning algorithms. The NN along with gradient-based optimization techniques is most popular choice among all researchers and associated with the short comings of local minima and sensitivity to initial value persists as a result of poor accuracy. So, as to resolve above said problems, global evolutionary algorithms (EA) such as Genetic Algorithms (GA) [1, 23], Particle Swarm Optimization (PSO) [19, 33, 34] have been utilized. The main advantages of EA lie in its global convergence, inherent parallel search nature, and great robustness. These algorithms generate a high quality solution within a short computation time.

For proper input selection, there is need of complete experimental analysis on the basis of error rate. The input structures of WT based models are different from that of the non WT based models. In the WT based models, the input is the combination of Wind Power series and WT based approximated and detailed wind power series. Therefore, the number of input nodes is more as compared to non WT models. The structure of WT based FFNN for wind power prediction has been shown in **Figure 2** & detailed prediction steps are:

Step 1: From the raw data of wind power, a time series as input is selected on the behalf of ACF.

Step 2: Supply the created input signal to WT for performing multilevel decomposition on wind power signal by utilizing Daubechies (db10) wavelet.

Step 3: Now extract the multi level approximation A6 and 1, 2, 3,4,5,6 level detailed coefficients D1 to D6 of input wind power series signal.

Step 5: The approximated and detailed wind power series along with six original time lags has been used as an input variables.

Step 6: A three layer FFNN, as shown in **Figure 3**, has been selected having thirteen input nodes equal to the number of input variables, twelve hidden neurons with tangential sigmoid transfer function, and one output neuron with pure linear activation function, with each series. The network is trained using Levenberg–Marquardt (LM) training algorithms with architecture [12–11–1]. The momentum constant and learning rate have been kept equal to 0.06 and 0.001, respectively.

Step 7: For the prediction, one year wind data has been trained and tested for next one month, similar process is continuously repeated up-to next 24 months with one month moving window. The maximum epochs were set equal to 10,000 with the performance goal of 0.001.

Step 8: The output values found by the network has been assessed on the accuracy criterion with actual wind power data series.

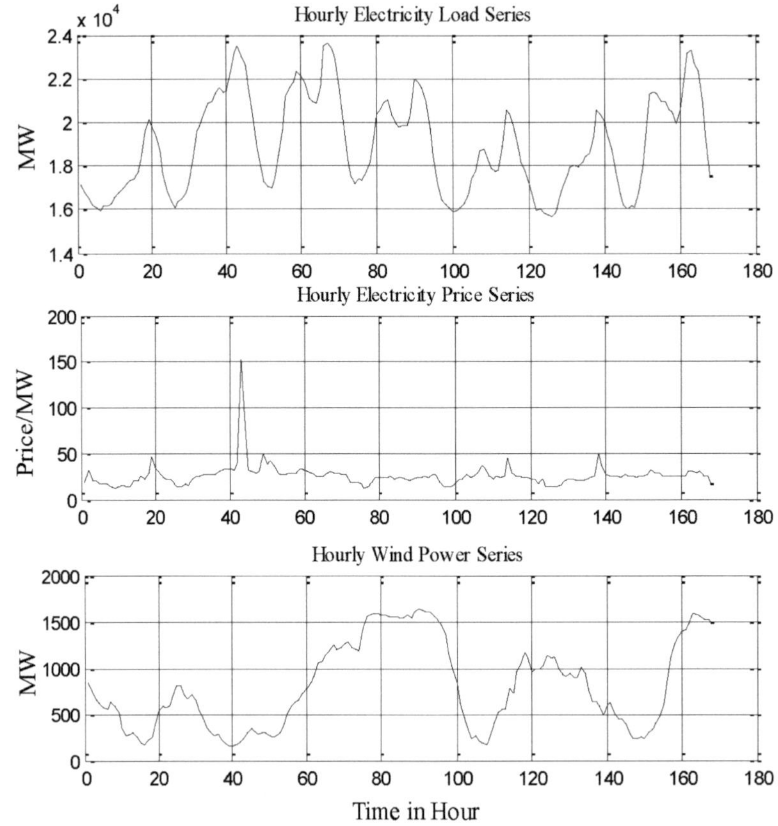

Figure 2.
Hourly curve for load, price & wind power from Ontario electricity market.

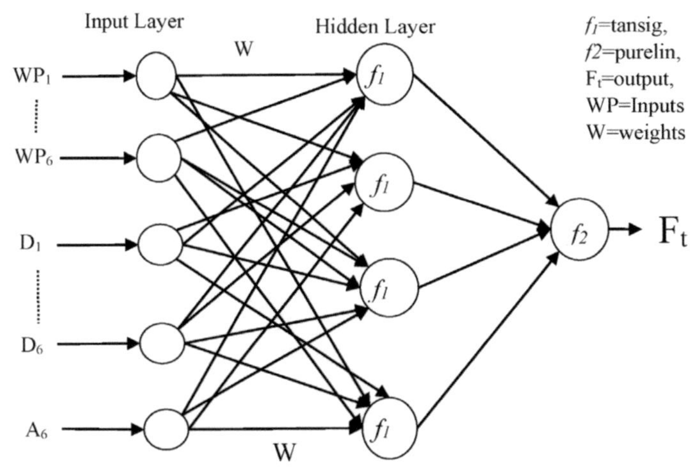

Figure 3.
WT based FFNN for wind power forecasting.

3.4 Evaluation of prediction performance

The aim of forecast evaluation is to assess, the general quality of a forecast by comparing the forecasted system states to actual observed states. The forecast evaluation provides a forecaster with:

- The ability of better improvement and understanding of forecast. The evaluation of forecast exposes all those sub-spaces whose forecasting error is more out of model state space. So, a forecaster can take advantage of analyzing sub-spaces & utilize it for improving forecasting model.

- Justifying the cost associated with resources used in forecasting model. The forecast performance assessment in accuracy terms gives a measure that can be directly linked to the utility or forecast user. Then coast and utility are compared with each other.

- The ability of performing model selection so that maximum certainty of results can be obtained with the comparison of others.

In most of the forecasting models accuracy is the criterion for selecting a particular method for the forecasting. For a consumer accuracy of forecasting is most important. The various methods for accuracy calculation given below:

- **The Error**

$$E = (WP_t - F_t) \tag{6}$$

Where, WP_t, is actual observation at time t, F_t, is forecast for time t

- **The Mean Absolute Error (MAE)**

$$MAE = \frac{1}{n}\sum_{t=1}^{n}|WP_t - F_t| \tag{7}$$

- **The Root Mean Square Error (RMSE)**

$$RMSE = \sqrt{\frac{1}{n}\sum_{t=0}^{n}(WP_t - F_t)^2} \tag{8}$$

- **Percentage Error (PE)**

$$PE = \left(\frac{(WP_t - F_t)}{Y_t}\right) * 100 \tag{9}$$

- **The Mean Absolute Percentage Error (MAPE)**

$$MAPE = \frac{1}{n}\sum_{n=1}^{t}|PE| \tag{10}$$

The prediction performance of forecasting carried out by the different models used in this research is justified on the basis of forecasting accuracy indices. The methodology described above has been applied to predict the wind power of OEM for two years from November 2012 to October 2014 on MAPE & MAE accuracy criteria. The software used for training and testing of NN is MATLAB version R2011b. The extensive use of WT for data pre-processing makes the results more significant and effective. From the results **Table 3**, it is clear that the results achieved with the help of WT based models have been found to be better up to 40–60% as compare to non WT based models. The 24 hours actual and forecasted wind power curves with error curve have been shown in **Figure 4**.

Model	Naïve	FFNN	ERNN	GANN	PSONN	GAPSONN	GRNN	LNNTD	WT + FFNN
MAPE	15.016	13.83	13.885	14.015	13.91	13.915	14.48	13.825	5.948
MAE	65.073	58.415	58.145	58.413	58.4675	58.29209	62.285	58.0475	23.225

Table 3.
Overall prediction comparisons for all models used.

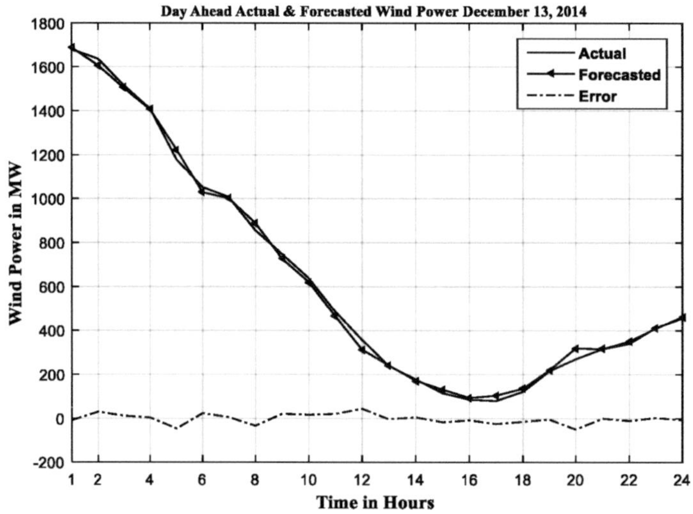

Figure 4.
One day ahead actual & forecasted wind power curve during winter season.

3.5 Uncertainty of forecasts using probabilistic forecasting

The uncertainty of forecasts is mainly due to the noise of training data, the misspecification of NN model for regression and input data selection.

NN Model Uncertainty: Uncertainty in NN forecasting arises due to misspecification in input parameters and structure of model which occurs due to local minima in the training process, random generation of input weights and so on. In case of global minima, misspecifications lead to non-eligible uncertainties in results related to prediction. The other factor behind model uncertainty is that during training finite samples never guarantee consistent generalization in performance of NN for future days. Basically, in WPF, it has become impossible to gather accurate information for reducing uncertainties while predicting and hence collectively called as model uncertainty. Due to model uncertainty, uncertainty in output should be handled carefully for accurate estimation in NN.

Data Uncertainty: Not only model uncertainty but also data noise adds to prediction uncertainty. If the data is stochastic in nature, then modeling is deterministically is really difficult. Both model misspecification and data noise are the major sources of uncertainties that affect the forecasting results.

In this, probabilistic forecasting of wind power has been performed in coordination with single step ahead wind power point forecasts. The major emphasis of probabilistic forecasting is to take into account the uncertainty associated with the wind power with probabilistic forecasting attributes such as: sharpness, reliability, resolution and discrimination. It consists of a set of prediction intervals which works in coordination with the best forecasts of single step ahead of wind power for the next coming hour; the interval forecasting has been incorporated. With a

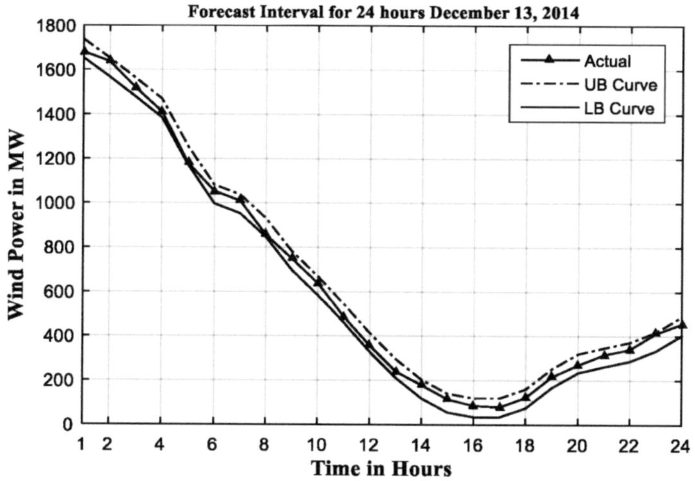

Figure 5.
PI with nominal confidence 95% in 24 hours look ahead.

pre-assumed probabilistic value, the basic aim of interval forecasting is to find out the range of prediction interval in which next hour wind power output lies. This framework has been consequently used for evaluating and analyzing the skill of the models for one lead hour point forecast. Thus, the overall results have been proving the reliability of results and show how the resolution may improve the forecasts skill.

The probabilistic forecasting has a wide range of statistical parameters on which the probabilistic outcomes of wind power lies. The prediction intervals (PI) stands for a wide range of possible probabilistic values within which the observed wind power values lies with a certain predefined probability. The basic idea behind the prediction intervals is to estimate the uncertainty associated with observed wind power $\left(WP_i^t\right)$ and forecasted $F\left(WP_i^t\right)$. The prediction intervals range can be much more enclosed and wider both depending on the value of confidence intervals (CI). The CI can be expressed as:

$$\text{Confidence Interval (CI)} = 100(1 - \alpha)\% \tag{11}$$

For a given sample size α has been a significant level which has been used to take into account the CI of the certain prediction intervals. The probabilistic stochastic interval (PSI) can be obtained by:

$$PSI_t^\alpha(WP_i) = \left[LB_t^\alpha(WP_i), UB_t^\alpha(WP_i)\right] \tag{12}$$

In the Eq. (12), the lower bound and upper bound can be expressed as:

$$LB_t^\alpha(WP_i) = F(WP_i) - z_{1-\sigma/2}\left(\frac{\sigma}{\sqrt{n}}\right) \tag{13}$$

$$UB_t^\alpha(WP_i) = F(WP_i) + z_{1-\sigma/2}\left(\frac{\sigma}{\sqrt{n}}\right) \tag{14}$$

In (13) and (14) $z_{1-\sigma/2}$ is the critical value of standard Gaussian distribution, which depends on certain value of CI, n is look ahead hour for the prediction sample & σ is the standard deviation of predicted values [11, 35–37] which is expressed as:

$$\sigma = \sqrt{\frac{\left(WP - \overline{WP}\right)^2}{(n-1)}} \tag{15}$$

For the WT based model, the upper bound curve and lower bound curves obtained at 95% of the confidence and the actual measured wind power curve in 24 hours has been shown in **Figure 5**.

4. Conclusions

The uncertainty, complexity and seasonal aspects associated with the wind contribute high level of uncertainties in wind power generation. Because weather conditions and wind speeds vary very much in different seasons. Therefore, for a perfect efficient forecasting model it is necessary to take care of input variables and their proper selection in time series. Actually, the improper input cause improper training of NN model as a result of that poor accuracy of forecasts. In this chapter, in order to take care of models forecast performance, probabilistic parameters have been taken into consideration.

In order to evaluate the performance on probabilistic forecasting, on the basis of single step reliable Prediction Intervals (PI's) need to be derived. In this, instead of exact values of forecast a range of forecasting interval need to be considered. If the predicted values lie in that range then, the performance of model is good otherwise model is poor one. Furthermore, power system operations require useful efficient forecast values with high level of reference confidence. Therefore, to fulfill the need of power system, more practical data based model should be required with high-confidence-level PI's.

Author details

Sumit Saroha[1*], Sanjeev Kumar Aggarwal[2] and Preeti Rana[3]

1 Assistant Professor, Department of Electrical Engineering, Guru Jambheshwar University of Science and Technology, Hisar, India

2 Associate Professor, Electrical and Instrumentation Engineering Department, Thapar Institute of Engineering and Technology, Patiala, Punjab, India

3 Scholar, MTH 48, Guru Jambheshwar University of Science and Technology, Hisar, India

*Address all correspondence to: saroha_sumit0178@yahoo.com

References

[1] Pai, P. F. and Hong, W. C.: Forecasting regional electricity load based on recurrent support vector machines with genetic algorithms. International Journal of Electric Power Systems Research. 2005;74:417–425. DOI: 10.1016/j.epsr.2005.01.006

[2] Kumar, A., Srivastva, S. C. and Singh, S. N.: Congestion management in competitive power market: A bibliography survey. International Journal of Electric Power Systems Research. 2005; 76:153–164. DOI: https://doi.org/10.1016/j.epsr.2005.05.001

[3] Kumar, A., Srivastva, S. C. and Singh, S. N.: A zonal congestion management approach using real and reactive power rescheduling. IEEE Transactions on Power Systems. 2004; 19(1): 554–562. DOI: 10.1109/TPWRS.2003.821448

[4] Negnevitsky, M., Mandal, P. Srivastava, A. K.: Machine learning applications for load, price and wind power prediction in power systems. 15th International Conference on Intelligent System Applications to Power Systems. 2009.

[5] Global Wind Energy Council Report. Available at: www.gwec.net, 2016.

[6] Aggarwal, S. K., Saini, L. M. and Kumar, A. (2009). Electricity price forecasting in deregulated markets: A review and evaluation, International Journal of Electrical Power and Energy Systems, 2009; 31: 13–22. DOI: https://doi.org/10.1016/j.ijepes.2008.09.003

[7] Amjady, N. and Hemmati, M. (2006). Energy price forecasting, IEEE Power & Energy Magazine, 2006; 20–29. DOI: 10.1109/MPAE.2006.1597990

[8] Wang, X., Guo, P. and Huang, X. (2011). A review of wind power forecasting models. Energy Procedia. 2011; 12: 770–778. DOI: https://doi.org/10.1016/j.egypro.2011.10.103

[9] Togelou, A., Sideratos, G. and Hatziargyriou, N. D.: Wind power forecasting in the absence of historical data. IEEE Transactions on Sustainable Energy. 2012; 3(3): 416–421. DOI: 10.1109/TSTE.2012.2188049

[10] Amjady, N. and Keynia, F.: Day-ahead price forecasting of electricity markets by mutual information technique and cascaded neuro-evolutionary algorithm, IEEE Transactions on Power Systems. 2009; 24(1): 306–318. DOI: 10.1109/TPWRS.2008.2006997

[11] Pinson, P. and Kariniotakis, G.: Conditional prediction intervals of wind power generation. IEEE Transactions on Power Systems. 2010; 25(4): 1845–1856. DOI: 10.1109/TPWRS.2010.2045774

[12] Aggarwal, S. K., Saini, L. M. and Kumar, A.: Day-ahead price forecasting in Ontario electricity market using variable-segmented support vector machine-based model. International Journal of Electric Power Components and Systems. 2009; 37: 495–516. DOI: https://doi.org/10.1080/15325000802599353

[13] Olsson, M., Perninge, M. and Soder, L.: Modeling real-time balancing power demands in wind power systems using stochastic differential equations. International Journal of Electric Power Systems Research. 2010; 80: 966–974. DOI: https://doi.org/10.1016/j.epsr.2010.01.004

[14] Ernst, B., Oakleaf, B., Ahlstrom, M. L., Lange, M., Moehrlen, C., Lange, B., Focken, U. and Rohrig, K.: Predicting the wind. IEEE Power & Energy Magazine. 2007; 78–89. DOI: 10.1109/MPE.2007.906306

[15] Zhao, P., Wang, J. F., Xia, J., Dai, Y., Sheng, Y. and Yue, J.: Performance

evaluation and accuracy enhancement of a day-ahead wind power forecasting system in China. International Journal of Renewable Energy. 2011; 43:234–241. DOI: https://doi.org/10.1016/j.renene.2011.11.051

[16] Ramirez-Rosado, I. J., Fernandez-Jimenez L. A., Monteiro, C., Sousa, J. and Bessa, R.: Comparison of two new short-term wind-power forecasting systems, International Journal of Renewable Energy, 2009; 34: 1848–1854. DOI: https://doi.org/10.1016/j.renene.2008.11.014

[17] Chen, K. and Yu, J.: Short-term wind speed prediction using an unscented kalman filter based state-space support vector regression approach, International Journal of Applied Energy, 2014; 113: 690–705. DOI: https://doi.org/10.1016/j.apenergy.2013.08.025

[18] Ontario electricity market data. Available at: http://www.ieso.ca/en/power-data/data-directory.

[19] Catalão, J. P. S., Pousinho, H. M. I., Mendes, V. M. F.: Hybrid wavelet-PSO-ANFIS approach for short-term wind power forecasting in Portugal, IEEE Transactions on Sustainable Energy, 2011; 2(1): 50–59. DOI: 10.1109/TSTE.2010.2076359

[20] Liu, H. and Tian, H. Q., Chen, C. and Li, Y. F.: A hybrid statistical method to predict wind speed and wind power, International Journal of Renewable Energy, 2010; 35(8): 1857–1861. DOI: https://doi.org/10.1016/j.renene.2009.12.011

[21] An, X., Jiang, D., Liu, C. and Zhao, M.: Wind farm power prediction based on wavelet decomposition and chaotic time series, International Journal of Expert Systems with Applications, 2011; 38(9): 11280–11285. DOI: https://doi.org/10.1016/j.eswa.2011.02.176

[22] Catalão, J. P. S., Pousinho, H. M. I., Mendes, V. M. F.: Short-term wind power forecasting in Portugal by neural networks and wavelet transform. International Journal of Renewable Energy. 2011; 36(4): 1245–1251. DOI: https://doi.org/10.1016/j.renene.2010.09.016

[23] Liu, D., Niu, D., Wang, H. and Fan, L.: Short-term wind speed forecasting using wavelet transform and support vector machines optimized by genetic algorithm. International Journal of Renewable Energy. 2014; 62: 592–597. DOI: https://doi.org/10.1016/j.renene.2013.08.011

[24] Bhaskar, K. and Singh, S. N.: AWNN-assisted wind power forecasting using feed-forward neural network. IEEE Transactions on Sustainable Energy. 2012; 3(2):306–315. DOI: 10.1109/TSTE.2011.2182215

[25] An, X., Jiang, D., Zhao, M. and Liu, C.: Short-term prediction of wind power using EMD and chaotic theory, International Journal of Communication Nonlinear Science Numerical Simulation, 2012; 17: 1036–1042. DOI: https://doi.org/10.1016/j.cnsns.2011.06.003

[26] Taieb, S. B., Gianluca, B., Atiya, A. F. and Sorjamaa, A.A.: Review and comparison of multi-step ahead time series forecasting based on the NN5 forecasting competition. International Journal of Expert Systems with Applications. 2012; 39: 7067–7083. DOI: https://doi.org/10.1016/j.eswa.2012.01.039

[27] Felice, M. D. and Yao, X.: Short-term load forecasting with neural network ensembles: A comparative study. IEEE Computational Intelligence Magazine. 2011; 47–56. DOI: 10.1109/MCI.2011.941590

[28] Aggarwal, S. K., Saini, L. M. and Kumar, A.: Short term price forecasting

in deregulated electricity markets-A review of statistical models and key issues. International Journal of Energy Sector Management. 2009; 3(4): 333–358. DOI: https://doi.org/10.1108/17506220911005731

[29] Kavasseri, R. G. and Seetharaman, K.: Day-ahead wind speed forecasting using f-ARIMA models, International Journal of Renewable Energy. 2009; 34: 1388–1393. DOI: 10.1016/j.renene.2008.09.006

[30] Erdem, E. and Shi, J.: ARMA based approaches for forecasting the tuple of wind speed and direction, International Journal of Applied Energy, 2011; 88(4): 1405–1414. DOI: 10.1016/j.apenergy.2010.10.031

[31] Hill, D. C., McMillan, D., Bell, K. R. W. and Infield, D.: Application of auto-regressive models to U.K. wind speed data for power system impact studies, IEEE Transactions on Sustainable Energy, 2012; 3(1):134–141. DOI: 10.1109/TSTE.2011.2163324

[32] Erdem, E. and Shi, J.: ARMA based approaches for forecasting the tuple of wind speed and direction, International Journal of Applied Energy, 2011; 88(4): 1405–1414. DOI:10.1016/j.apenergy.2010.10.031

[33] Bashir, Z. A. and El-Hawary, M. E.: Applying wavelets to short-term load forecasting using PSO-based neural networks, IEEE Transactions on Power Systems, 2009; 24(1): 20–27. DOI: 10.1109/TPWRS.2008.2008606

[34] Yang, X., Yuan, J., Yuan, J. and Mao, H.: An improved WM method based on PSO for electric load forecasting, International Journal of Expert Systems with Applications, 2010; 37:8036–8041. DOI: https://doi.org/10.1016/j.eswa.2010.05.085

[35] Sideratos, G. and Hatziargyriou, N. D.: Probabilistic wind power forecasting using radial basis function neural networks, IEEE Transactions on Power Systems, 2012; 27(4): 1788–1796. DOI: 10.1109/TPWRS.2012.2187803

[36] Sideratos, G. and Hatziargyriou, N. D.: An advanced statistical method for wind power forecasting, IEEE Transactions on Power Systems, 2007; 22(1): 258–265. DOI: 10.1109/TPWRS.2006.889078

[37] Gneiting, T., Balabdaoui, F. and Raftery, A. E.: Probabilistic forecasts, calibration and sharpness, Journal of Royal Statistical Society, 2007; 69(2): 243–268. DOI: https://doi.org/10.1111/j.1467-9868.2007.00587.x

Chapter 5

Stock Market Trend Prediction Using Hidden Markov Model

Deneshkumar Venugopal,
Senthamarai Kannan Kaliyaperumal
and Sonai Muthu Niraikulathan

Abstract

In Recent years many forecasting methods have been proposed and implemented for the stock market trend prediction. In this Chapter, the trend analyses of the stock market prediction are presented by using Hidden Markov Model with the one day difference in close value for a particular period. The probability values π gives the trend percentage of the stock prices which is calculated for all the observe sequence and hidden sequences. This chapter helps for decision makers to make decisions in case of uncertainty on the basis of the percentage of probability values obtained from the steady state probability distribution.

Keywords: stock market, HMM, TPM, EPM and trend prediction

1. Introduction

The fundamental idea behind a hidden Markov model is that there is a Markov process we cannot observe that determines the probability distribution for what we do observe. Thus a hidden Markov model is specified by the transition density of the Markov chain and the probability laws that govern what we observe given the state of the Markov chain. Given such a model, we want to estimate any parameters that occur in the model. And also determined the most likely sequence for the hidden process. Finally we may want the probability distribution for the hidden states at every location.

Let y_t represents the observed value of the process at location t for $t = 1, \ldots, T$, θ_t the value of the hidden process at location t and let ϕ represents parameters necessary to determine the probability distribution for y_t given θ_t and θ_t given θ_{t-1}. In our applications, y_t will either be an increase or decrease and the hidden process will determine the probability distribution of observing different letters.

Our model is then described by the sets of probability distributions $p(y_t | \theta_t, \phi)$ and $p(\theta_t | \theta_{t-1}, \phi)$. A crucial component of this model is that the y_t are independent given the set of θ_t and θ only depends directly on its neighbors θ_{t-1} and θ_{t+1}. The various distribution in which we are interested are $p(\phi | y_1, \ldots, y_T)$, $p(\theta_t | y_1, \ldots, y_T)$ for all t and $p(\theta_1, \ldots, \theta_T | y_1, \ldots, y_t)$. We will adopt a Bayesian perspective, so that we treat θ_t as a random variable [1, 2].

The measure of best is to find the path that has the maximum probability in the HMM, given the sequence X. Recall that the model gives the joint probabilities

$Pr(H, X)$ for all sequence, it also gives the posterior probability $Pr(H, X) = Pr(H, X)/Pr(X)$, for every possible state path H through the model, conditioned on the sequence X with maximum posterior probability [3, 4]. Given that the denominator $Pr(X)$ is constant in the conditional probability formula for a given sequence X, maximizing the posterior probability is equivalent to finding the state path H* that maximizes the joint probability $Pr(H^*, X)$. Nguyen [5] has determined the optimal number of states for the HMM by using the AIC, BIC and HQ information criteria and also discussed the applications of HMM in stock trading. Hassan and Nath [6] have applied HMM to the airlines stock forecast. HMMs have been used for pattern recognition and classification problems and it was suitable for modeling dynamic systems.

2. Hidden Markov model

Hidden Markov model (HMM) is a stochastic model which is not directly observable, It describes the observable events that are depends on internal factors. The observable events are represented as symbols, where the invisible factor involved in the observation is represented as a state. HMM is a stochastic model where the system is assumed to be a Markov Process with hidden states and it gives better accuracy than the other models. Using the given input values, the parameters of the HMM (λ) denoted by A, B and π are found out. An HMM is defined as λ = (S,O,A,B,π) where S = {s1,s2, … ,sN} is a set of N possible states O = {o1,o2, … ,oM} is a set of M possible observation symbols, A is an N*N state Transition Probability Matrix (TPM), B is an N*M observation or Emission Probability Matrix (EPM) and Π is an N dimensional initial state probability distribution vector and A,B and π should satisfy the following conditions (**Figure 1**):

$$\sum_{j=1}^{N} a_{ij} = 1 \, where \, 1 \le i \le N;$$

$$\sum_{j=1}^{M} b_{ij} = 1 \, where \, 1 \le i \le N;$$

$$\sum_{i=1}^{N} \pi_i = 1 \, where \, \pi_i \ge 0$$

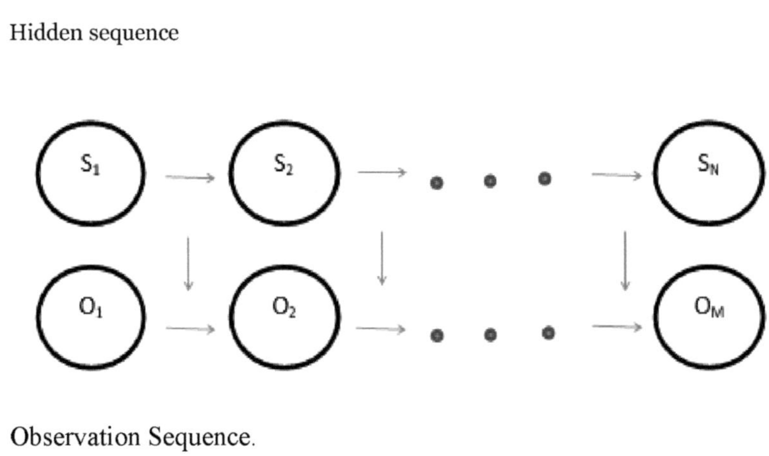

Figure 1.
Diagram of HMM.

2.1 Evaluation problem

Given the HMM = {A,B,π} and the observation sequence O = o1,o2, ... ,oM, the probability that model λ has generated sequence O is calculated. Often this problem is solved by the Forward Backward Algorithm [7, 8].

2.2 Decoding problem

Given the HMM λ = {A,B,π} and the observation sequence O = o1,o2, ... ,oM, calculate the most likely sequence of hidden states that produced this observation sequence O. Usually this problem is handled by Viterbi Algorithm [7, 8].

2.3 Learning problem

Given some training observation sequences O = o1,o2, ... ,oM, and general structure of HMM (numbers of hidden and visible states), determine HMM parameters λ = {A,B,π} that best fit training data. The most common solution for this problem is Baum-Welch algorithm [9, 10] which is considered as the traditional method for training HMM.

3. Results and discussions

In this chapter, the data has been taken from Yahoofinance.com and the NSE daily close value data for a month of January 2020 period is considered for the analysis.

Here two observing symbols “I” for Increasing states and the symbols “D” for decreasing states have been used. If the differences of close value greater than 0 its observing that the symbol is “f” and If the differences of close value less than 0 its observing that the symbol is “D”. There are six hidden states assumed and are denoted by the symbol S1, S2, S3, S4, S5, S6 are indicates that very low, low, moderate low, moderate high, high and very high respectively. The states are not directly observable.

The situations of the stock market are considered hidden. Given a sequence of observation we can find the hidden state sequence that produced those observations. **Table 1** shows the daily close value of the stock market.

Interval values:
S1 = −9500 to −551.
S2 = −550 to −251.
S3 = −250 to 249.
S4 = 250 to 8500.

The various probability values of TPM, EPM and π for difference in one day, two days, three days, four days, five days, six days close value are calculated as given below (**Table 2**).

Probability values of TPM, EPM, and π for difference in one day close value (**Figure 2** and **Table 3**):

$$\begin{bmatrix} & S1 & S2 & S3 & S4 \\ S1 & 0 & 0 & 1 & 0 \\ S2 & 0 & 0 & 1 & 0 \\ S3 & 0.071 & 0.071 & 0.4286 & 0.4286 \\ S4 & 0 & 0 & 1 & 0 \end{bmatrix} \quad \begin{bmatrix} & I & D \\ S1 & 0 & 1 \\ S2 & 0 & 1 \\ S3 & 0.2849 & 0.7143 \\ S4 & 0.5 & 0.5 \end{bmatrix}$$

S. no	Date	Close
1	01/02/2020	41,626.64
2	01/03/2020	41,464.61
3	01/06/2020	40,676.63
4	01/07/2020	40,869.47
5	01/08/2020	40,817.74
6	01/09/2020	41,452.35
7	01/10/2020	41,599.72
8	01/13/2020	41,859.69
9	01/14/2020	41,952.63
10	01/15/2020	41,872.73
11	01/16/2020	41,932.56
12	01/17/2020	41,945.37
13	01/20/2020	41,528.91
14	01/21/2020	41,323.81
15	01/22/2020	41,115.38
16	01/23/2020	41,386.4
17	01/24/2020	41,613.19
18	01/27/2020	41,155.12
19	01/28/2020	40,966.86
20	01/29/2020	41,198.66
21	01/30/2020	40,913.82
22	01/31/2020	40,723.49

Table 1.
Daily close value of NSE.

Probability values of TPM, EPM, and π for difference in two day close value (**Figure 3** and **Table 4**).

$$\begin{bmatrix} & S1 & S2 & S3 & S4 \\ S1 & 0 & 0 & 1 & 0 \\ S2 & 0 & 0 & 1 & 0 \\ S3 & 0.0111 & 0 & 0.5555 & 0.3333 \\ S4 & 0 & 0.3333 & 0.5 & 0.1667 \end{bmatrix} \quad \begin{bmatrix} & I & D \\ S1 & 0.5 & 0.5 \\ S2 & 0.5 & 0.5 \\ S3 & 0.4444 & 0.5556 \\ S4 & 1 & 0 \end{bmatrix}$$

Probability values of TPM, EPM, and π for difference in three day close value (**Figure 4** and **Table 5**):

$$\begin{bmatrix} & S1 & S2 & S3 & S4 \\ S1 & 0 & 0 & 0 & 1 \\ S2 & 0 & 0 & 0.75 & 0.25 \\ S3 & 0 & 0.6 & 0.2 & 0.2 \\ S4 & 0.5 & 0.2 & 0.2 & 0.2 \end{bmatrix} \quad \begin{bmatrix} & I & D \\ S1 & 0 & 1 \\ S2 & 1 & 1 \\ S3 & 0.6 & 0.4 \\ S4 & 1 & 0 \end{bmatrix}$$

S. no	c.v	D in 1 day CV	o.s	D in 2 days CV	o.s	D in3 days CV	o.s	D in 4 day CV	o.s	D in 5 day CV	o.s	D in6 days CV	o.s
1	41,626.64												
2	41,464.61	162.03	I										
3	40,676.63	787.98	I	−625.95	D								
	40,869.47	−192.84	D	980.82	I	−1606.77	D						
5	40,817.74	51.73	I	−244.57	D	1225.39	I	−2882.16	D				
6	41,452.35	−634.61	D	686.84	I	−930.91	D	2156.3	I	−4988.46	D		
7	41,599.72	−147.37	D	−487.24	D	1173.58	I	2104.49	I	4260.79	I	−9249.25	D
8	41,759.69	−259.97	D	112.6	I	−599.84	D	1773.42	I	−3877.91	D	8138.7	I
9	41,952.63	−92.94	D	−167.03	D	279.63	I	−879.47	D	2652.89	I	−6530.8	D
10	41,872.73	79.9	I	−172.84	D	5.81	I	273.82	I	−1153.28	D	3806.18	I
11	41,932.56	−59.83	D	139.73	I	−312.57	D	318.38	I	−44.56	D	−1108.73	D
12	41,945.37	−12.81	D	−47.02	D	−92.71	D	405.28	I	−86.9	D	42.34	I
13	41,528.91	416.46	I	403.65	I	−450.67	D	357.96	I	47.32	I	−134.22	D
14	41,323.81	205.1	I	211.36	I	192.22	I	−642.96	D	1000.92	I	−953.6	D
15	41,115.38	208.43	I	−3.33	D	214.69	I	−22.4	D	−620.56	D	1621.48	I
16	41,386.4	−271.02	D	479.45	I	−482.78	D	697.47	I	−719.87	D	99.31	I
17	41,613.19	−226.79	D	−44.23	D	523.68	I	−1006.46	D	1703.93	I	−2423.8	D
18	41,155.12	458.01	I	−684.86	D	640.63	I	−116.95	D	−889.51	D	2593.44	I
19	40,966.86	188.26	I	269.81	I	−415.05	D	1055.68	I	938.73	I	−1828.24	D
20	41,198.66	−231.8	D	420.06	I	−150.25	D	−264.8	D	1320.48	I	−381.75	D
21	40,913.82	284.84	I	−516.64	I	936.7	I	−1086.95	D	822.15	I	498.33	I
22	40,723.49	190.33	I	94.51	I	−611.15	D	1547.85	I	−2634.8	D	3456.95	I

Table 2.
Daily close value for finding differences in one day, two day, three days, four days, five days, six days close value.

	S1		S2		S3		S4	
	I	**D**	**I**	**D**	**I**	**D**	**I**	**D**
S1	0	0	0	0	1	0	0	0
S2	0	0	0	0	1	0	0	0
S3	0.071	0	0.071	0	0.1429	0.2857	0	0.4286
S4	0	0	0	0.8	0.2	0	0	0

Table 3.
Transitions with probability values for one day close value.

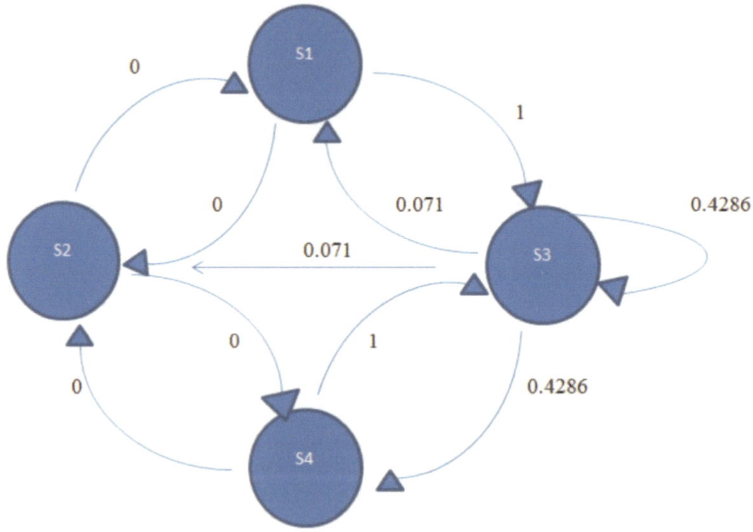

Figure 2.
Diagram of TPM day 1.

	S1		S2		S3		S4	
	I	**D**	**I**	**D**	**I**	**D**	**I**	**D**
S1	0	0	0	0	0	0	0.5	0.5
S2	0	0	0	0	0.5	0.5	0	0
S3	0	0.111	0	0	0.3333	0.2222	0.1111	0.2222
S4	0	0	0.3333	0	0.5	0	0.1667	0

Table 4.
Transition table with probability values for difference in two day close value.

Probability values of TPM, EPM and π for difference in four days close value (**Figure 5** and **Table 6**):

$$\begin{bmatrix} & S1 & S2 & S3 & S4 \\ S1 & 0.3858 & 0.1429 & 0.1429 & 0.4286 \\ S2 & 0.5 & 0 & 0.5 & 0 \\ S3 & 0 & 0 & 1 & 0 \\ S4 & 0.4286 & 0.1429 & 0 & 0.4286 \end{bmatrix} \quad \begin{bmatrix} & I & D \\ S1 & 0.1429 & 0.9573 \\ S2 & 0.5 & 0.5 \\ S3 & 0 & 1 \\ S4 & 1 & 0 \end{bmatrix}$$

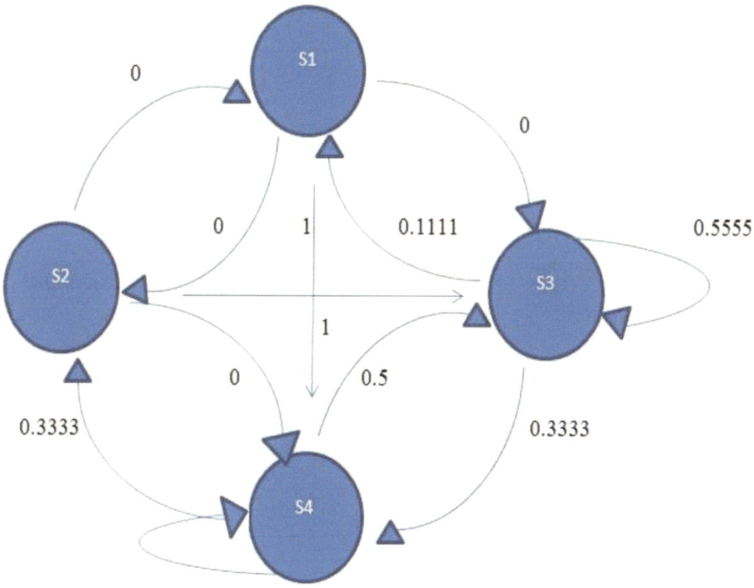

Figure 3.
Diagram of TPM day 2.

	S1		S2		S3		S4	
	I	**D**	**I**	**D**	**I**	**D**	**I**	**D**
S1	0	0	0	0	0	0	0	1
S2	0	0	0	0	0	0.75	0	0.25
S3	0	0	0.4	0.2	0.2	0	0	0.2
S4	0.5	0	0.2	0	0.2	0	0.2	0

Table 5.
Transition table with probability values for difference in three day close value.

Probability values of TPM, EPM and π for difference in five days close value (**Figure 6** and **Table 7**):

$$\begin{bmatrix} & S1 & S2 & S3 & S4 \\ S1 & 0.1667 & 0 & 0.1667 & 0.6667 \\ S2 & 0 & 0 & 0 & 0 \\ S3 & 0 & 0 & 0.6667 & 0.3333 \\ S4 & 0.7143 & 0 & 0 & 0.6667 \end{bmatrix} \quad \begin{bmatrix} & I & D \\ S1 & 0 & 1 \\ S2 & 0 & 1 \\ S3 & 0.3333 & 0.6667 \\ S4 & 1 & 0 \end{bmatrix}$$

Probability values of TPM, EPM and π for difference in six days close value (**Figure 7** and **Table 8**):

$$\begin{bmatrix} & S1 & S2 & S3 & S4 \\ S1 & 0 & 0.2 & 0.2 & 0.6 \\ S2 & 0 & 0 & 0 & 1 \\ S3 & 0.6667 & 0 & 0.3333 & 0 \\ S4 & 0.5 & 0 & 0.25 & 0.25 \end{bmatrix} \quad \begin{bmatrix} & I & D \\ S1 & 0 & 1 \\ S2 & 0 & 1 \\ S3 & 0.667 & 0.3333 \\ S4 & 1 & 0 \end{bmatrix}$$

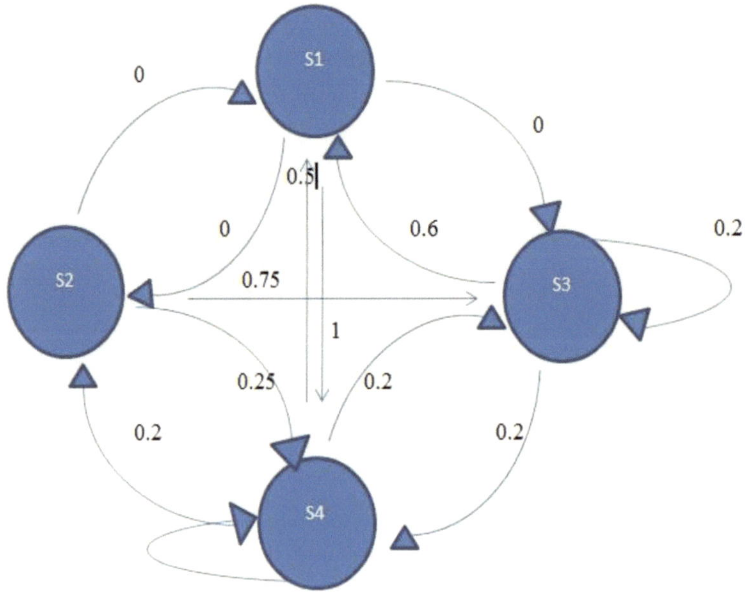

Figure 4.
Diagram of TPM day 3.

	S1		S2		S3		S4	
	I	D	I	D	I	D	I	D
S1	0.1429	0.2429	0	0.1429	0	0.1429	0	0.4286
S2	0.5	0	0	0	0	0.5	0	0
S3	0	0	0	0	0	1	0	0
S4	0.4286	0	0.1429	0	0	0	0.4286	0

Table 6.
Transition table with probability values for difference in four day close value.

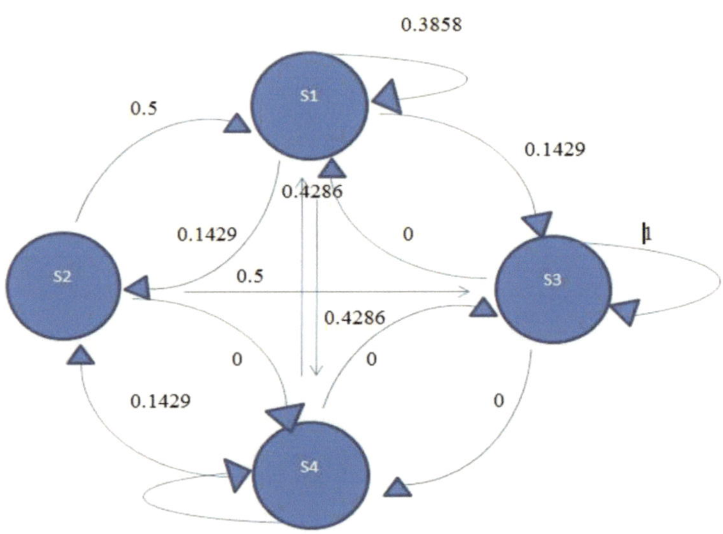

Figure 5.
Diagram of TPM day 4.

	S1		S2		S3		S4	
	I	**D**	**I**	**D**	**I**	**D**	**I**	**D**
S1	0	0.1667	0	0	0	0.1667	0	0.6667
S2	0	0	0	0	0	0	0	0
S3	0	0	0	0	0	0.6667	0.3333	0
S4	0.7143	0	0	0	0	0	0.2857	0

Table 7.
Transition table with probability values for difference in five day close value.

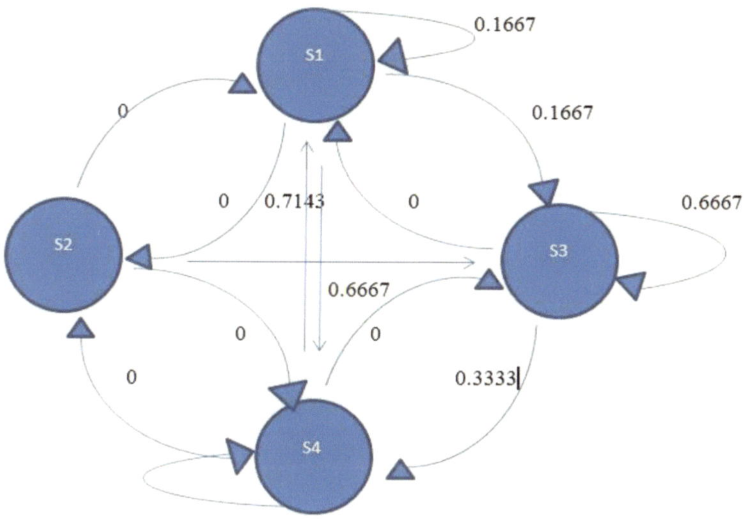

Figure 6.
Diagram of TPM day 5.

	S1		S2		S3		S4	
	I	**D**	**I**	**D**	**I**	**D**	**I**	**D**
S1	0	0	0	0.2	0	0.2	0	0.6
S2	0	0	0	0	0	0	0	1
S3	0.3333	0.3333	0	0	0.3333	0	0	0
S4	0.5	0	0	0	0.25	0	0.25	0

Table 8.
Transition table with probability values for difference in six day close value.

The various transitions probability values for difference in one day to six days close values are displayed in **Figure 2** to **Figure 7** respectively.

Optimum Sequence of States:

To generate a random sequence of emission symbols and states are calculated by using the function "Hmmgenerate". The HMM matlab toolbox syntax is: [Sequence, States] = Hmmgenerate(L,TPM,EPM). The length of both sequence and state to be generated is denoted by L [11]. The fitness function used for finding the fitted value of sequence of states is defined by

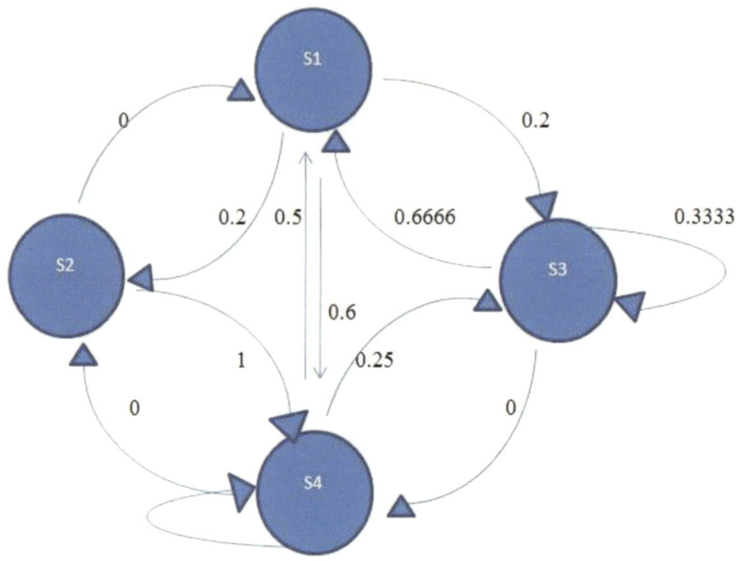

Figure 7.
Diagram of TPM day 6.

$$(\text{Fitness} =)\frac{1}{\sum compare(i,j)}$$

Using the iterative procedure, for each TPM and EPM framed we get an optimum sequence of states generated.

The length of the sequence taken as L = 4 and the optimum sequence of states obtained from the all six day's differences with TPM and EPM is given in the below and here 'ε' is the start symbol.

1.	ε	→	I S4	→	D S4	→	I S3	→	D S4
2.	ε	→	D S1	→	I S4	→	I S4	→	D S3
3.	ε	→	I S4	→	I s2	→	D S3	→	D S1
4.	ε	→	D S1	→	D S4	→	I S3	→	D S4
5.	ε	→	I S3	→	I S3	→	D S2	→	D S4
6.	ε	→	I S4	→	D S1	→	I S3	→	D S4

Here, the one day difference of TPM and EPM has the shortest path. So the best optimum sequence is found from one day difference in close value. Using the fitness function we compute the fitness value for each of the optimum sequences of states obtained (**Table 9**).

In column four the highest value is the fitness value and the better is the performance of the particular sequence.

S. no.	Comparison of six optimum sequence of states	Calculated value	Fitness = $\frac{1}{\sum comparision(i,j)}$
1	(1,2) + (1,3) + (1,4)	1	1
2	(2,1) + (2,3) + (2,4)	1.7	0.588
3	(3,1) + (3,2) + (3,4)	2.425	0.412
4	(4,1) + (4,2) + (4,3)	3.15	0.32

Table 9.
Comparison of six optimum state sequences.

4. Conclusion

Stock prediction is challenging due to its randomness. Hidden Markov Model can be used for stock prediction by finding hidden patterns. Here the Hidden Markov model easily recognized four states of the stock market and also it was used to predict the future values. The highest value in the Optimum State Sequences is the better performance of the particular sequence. Hidden states and sequences have been generated to easily identify the level of the sequence whether the next day value is increasing. And also identified whether the increasing level is moderate high or high or very high and also decreasing level whether moderate low or low or very low. This model will be very much useful for short term as well as long term investors.

Author details

Deneshkumar Venugopal*, Senthamarai Kannan Kaliyaperumal and Sonai Muthu Niraikulathan
Department of Statistics, Manonmaniam Sundaranar University, Tirunelveli, Abishekapatti, Tamil Nadu, India

*Address all correspondence to: vdenesh@msuniv.ac.in

References

[1] Medhi J. Stochastic processes. New Age International; 1994.

[2] Reilly C. Statistics in human genetics and molecular biology. CRC Press; 2009 Jun 19.

[3] Brejová B, Brown DG, Vinař T. ADVANCES IN HIDDEN MARKOV MODELS FOR SEQUENCE. Bioinformatics Algorithms: Techniques and Applications. 2008 Feb 15;3:55.

[4] Gupta A, Dhingra B. Stock market prediction using hidden Markov models. In 2012 Students Conference on Engineering and Systems 2012 Mar 16 (pp. 1-4). IEEE.

[5] Nguyen N. Hidden Markov model for stock trading. International Journal of Financial Studies. 2018 Jun;6(2):36.

[6] Hassan MR, Nath B. Stock market forecasting using hidden Markov model: a new approach. In 5th International Conference on Intelligent Systems Design and Applications (ISDA'05) 2005 Sep 8 (pp. 192-196). IEEE.

[7] Rabiner L. Theory and implementation of hidden Markov models. Fundamentals of speech recognition. 1993.

[8] Rabiner LR. A tutorial on hidden Markov models and selected applications in speech recognition. Proceedings of the IEEE. 1989 Feb;77(2): 257-286.

[9] Lloyd RW. Hidden Markov Models and the Baum-Welch Algorithm. IEEE Information Theory Society Newsletter. 2003 Dec;53(4).

[10] Mandoiu I, Zelikovsky A. Bioinformatics algorithms: techniques and applications. John Wiley & Sons; 2008 Feb 25.

[11] Murphy K. HMM toolbox for Matlab. Internet: http://www.cs.ubc. ca/ murphyk/Software/HMM/hmm. html, [Oct. 29, 2011]. 1998.

Chapter 6

Electric Load Forecasting an Application of Cluster Models Based on Double Seasonal Pattern Time Series Analysis

Ismit Mado

Abstract

Electricity consumption always changes according to need. This pattern deserves serious attention. Where the electric power generation must be balanced with the demand for electric power on the load side. It is necessary to predict and classify loads to maintain reliable power generation stability. This research proposes a method of forecasting electric loads with double seasonal patterns and classifies electric loads as a cluster group. Double seasonal pattern forecasting fits perfectly with fluctuating loads. Meanwhile, the load cluster pattern is intended to classify seasonal trends in a certain period. The first objective of this research is to propose DSARIMA to predict electric load. Furthermore, the results of the load prediction are used as electrical load clustering data through a descriptive analytical approach. The best model DSARIMA forecasting is ([1, 2, 5, 6, 7, 11, 16, 18, 35, 46], 1, [1, 3, 13, 21, 27, 46]) $(1, 1, 1)^{48}$ $(0, 0, 1)^{336}$ with a MAPE of 1.56 percent. The cluster pattern consists of four groups with a range of intervals between the minimum and maximum data values divided by the quartile. The presentation of this research data is based on data on the consumption of electricity loads every half hour at the Generating Unit, the National Electricity Company in Gresik City, Indonesia.

Keywords: electric loads, DSARIMA model, descriptive analytic, clustering, forecasting, time series

1. Introduction

Fluctuations in electrical power greatly affect the performance of power generation systems. Changes in electrical power due to variations in demand for electrical power momentarily result in an imbalance of electricity generated by the electric power absorbed. If the power supplied is greater than there will be energy waste. And if the power supplied is smaller then there will be overload which will result in a blackout. This means that the amount of electric power generated must be balanced or not too far from the nominal value of the electrical power requirements at the load center. In fact, the use of electrical energy tends to change at any time.

For this reason, it is necessary to predict the use of electric power that is able to maintain a balance between supply and consumption of electric power in the power generation system. Research of electricity load forecasting is very important in the power plant system operation plan [1]. Load forecasting studies are classified into three categories: long-term, medium-term and short-term predictions. Long-term predictions are needed for planning the peak load capacity and system maintenance schedule [2], medium-term predictions are needed for the planning and operation of the power plant system [3], and short-term predictions are needed to control and schedule the generating system [4]. So that load forecasting studies play a role in ensuring the economic value of financing, system reliability, stability and quality of electricity system services.

Fluctuations in electrical power at the load center contain a set of time-based information. The characteristics of the load from the period of use both by household, commercial, industrial and public costs, are needed so that fluctuations can be analyzed. The load characteristics, besides being able to be analyzed also contain a series of load patterns tendencies due to usage. This conduct of using electric loads contains seasonal patterns. Daily use tends to recur on certain days, as well as weekly load patterns. This trend is then analyzed through the load cluster approach to achieve load usage patterns based on seasonal patterns.

The Box-Jenkins time series study approach conducted in this research was able to increase the estimated usage and application of seasonal patterns based on electricity load clusters. The time series prediction model is an accurate choice and continues to grow to this day [5–7]. Researchers have carried out load forecasting study activities with 2.06 percent MAPE [8]. In research, the parameter estimation pattern was developed again with the least squares method which is better. And then the load cluster modeling is developed to classify the trend based on seasonal patterns.

2. Electrical load characteristics

The main purpose of an electric power distribution system is to distribute electric power from substations or sources to a number of customers or loads. The most important main factor in the distribution system planning is the characteristics of various electrical loads.

The electrical load characteristics are needed so that the system voltage, the thermal effect of loading and the loading pattern can be analyzed properly. The analysis is included in determining the initial projections in the next planning.

The characteristics of the electrical load are very dependent on the type of load it serves. This will be clearly seen from the results of recording the load curve in a time interval. The following are several factors that determine the load characteristics according to the needs of this study [9].

2.1 Load factor

Load factor is the ratio between average load and peak load measured in a certain period. Average load and peak load can be expressed in KiloWatt (KW), KiloVolt-Ampere (KVA) and so on, but the units of both must be the same. Load factor can be calculated for a certain period usually used in units of daily, monthly or yearly.

The peak load referred to in this study is a momentary peak load or average peak load in a certain interval (maximum demand), generally a maximum demand of

15 minutes or 30 minutes is used. In this study, the load data used is 30-minute interval load data.

The definition of the load factor can be written in the following equation:

when you are citing sources, the citations should be set in numbered format. All the references given in the list of references should be cited in the body of the text. Please set citations in square brackets keeping the below points in mind.

$$Load\ factor = \frac{average\ load\ in\ a\ certain\ period}{peak\ load\ in\ a\ certain\ period} \tag{1}$$

The load factor can be known from the load curve. As for the estimation of the magnitude of the burden factor in the future, it can be approached with existing statistical data as was done in this study.

When applied to the power plant, it is formulated into

$$Load\ factor = \frac{P_{average}}{P_{peak}} \times \frac{T}{T} \tag{2}$$

If T is in a year, an annual expense factor is obtained. If in 1 month the monthly load factor is obtained, as well as the daily load factor.

2.2 Daily load

Daily load factors vary according to the characteristics of the load area, whether it is a dense residential area, industrial area, trade or a combination of various types of customers.

This daily load factor will also affect the weather conditions and certain days such as holidays and so on.

2.3 Load curve

Load curves illustrate the variation of loading on a substation measured by KW or KVA as a function of time. Measurement time intervals are usually determined based on the use of measurement results, for example intervals of 30 minutes, 60 minutes, 1 day or 1 week.

The load curve shows the demand or load requirements at different time intervals. With the help of this load curve, we can determine the magnitude of the largest load and then the generating capacity can also be determined.

2.4 Peak load

Peak load or maximum demand is defined as the biggest load of needs that occurs during a certain period. Certain periods can be in the form of daily, monthly or annual periods. Furthermore, the peak load must be interpreted as the average load during a certain interval, where the possibility of such load. For example, the daily load of a distribution transformer where the peak load during an interval of 1 hour, ie between 19:00 (point A) and 20:00 (point B). The average value of the A - B curve is its peak requirement.

Keep in mind here that peak needs are not instantaneous needs, but on average during a certain time interval, usually a certain time interval is 15 minutes, 30 minutes or 1 hour.

The characteristics of the burden between holidays are different from ordinary days so that they have different load variants. Load characteristics can also be distinguished by the factor of loading outside the time of the peak load, or who are at the time of the peak load. So we need load forecasting with the aim of preparing operating generating units. When electricity demand increases, it will be balanced with adequate electricity supply to prevent power outages, otherwise if electricity consumption decreases, electricity supply will be reduced so as not to over supply.

3. Electrical load analysis based on time series model

Box and Jenkins popularized the use of ARIMA models and the Box-Jenkins methodology became highly popular in the 1970s among academics [10]. The ARIMA model is also called the Box-Jenkins time series. A time series is a series of observations taken sequentially based on time [11]. The observation process is carried out at the same interval, for example in hour, daily, weekly, monthly, yearly or other intervals. The purpose of time series analysis is twofold, namely to model the stochastic mechanism found in observations based on time and to predict the value of observations in the future. The value of a variable can be predicted if the nature of the variable is known in the present and in the past.

3.1 ARIMA model classification

The ARIMA model is divided into several groups, namely: autoregressive (AR), moving average (MA), and ARMA. The ARIMA model is a nonstationary ARMA model that has gone through a differencing process so that it becomes a stationary model. The ARIMA model also contains seasonal patterns. Defined as a pattern that repeats in a fixed time interval. The application of this seasonal pattern has been developed into a double seasonal pattern [12–14]. Double seasonal ARIMA model is written with notation, as follows.

$$ARIMA\ (p,d,q)(P_1,D_1,Q_1)^{S_1}(P_2,D_2,Q_2)^{S_2} \tag{3}$$

This model consists of two components, namely the first level which is usually developed from a linear forecasting model to explain seasonal trends from data or known as potential load. And at the second level developed from the ARIMA model to capture autoregressive patterns from data or called irregular loads. For stationary data, the seasonal factor can be determined by identifying the coefficient of autocorrelation at two or three time intervals that are very different from zero. So that this seasonal pattern can be identified whether it contains a tendency to have a seasonal pattern or multiple seasonal patterns and has the following general form [15]:

$$\phi_p(B)\Phi_{P_1}(B^{s_1})\Phi_{P_2}(B^{s_2})(1-B)^d(1-B^{s_1})^{D_1}(1-B^{s_2})^{D_2}Z_t = \theta_q(B)\Theta_{Q_1}(B^{s_1})\Theta_{Q_2}(B^{s_2})a_t \tag{4}$$

With

$\phi_p(B) = 1 - \phi_1 B - \phi_2 B^2 - \ \dots\ - \phi_p B^p$

$\Phi_{P_1}(B^{s_1}) = 1 - \Phi_{1_1}B^{s_1} - \Phi_{2_1}B^{2s_1} - \ \dots\ - \Phi_{P_1}B^{P_1 s_1}$

$\Phi_{P_2}(B^{s_2}) = 1 - \Pi_{1_2}B^{s_2} - \Pi_{2_2}B^{2s_2} - \ \dots\ - \Pi_{P_2}B^{P_2 s_2}$

$\theta_q(B) = 1 - \theta_1 B - \theta_2 B^2 - \ \dots\ - \theta_q B^q$

$\Theta_{Q_1}(B^{s_1}) = 1 - \Theta_{1_1}B^{s_1} - \Theta_{2_1}B^{2s_1} - \ \dots\ - \Theta_{Q_1}B^{Q_1 s_1}$

$\Theta_{Q_2}(B^{s_2}) = 1 - \Psi_{1_2}B^{s_2} - \Psi_{2_2}B^{2s_2} - \ \dots\ - \Psi_{Q_2}B^{Q_2 s_2}.$

3.2 ARIMA Box-Jenkins procedure

The prediction procedure of ARIMA Box-Jenkins model through five stages of iteration, as follows:

i. Preparation of data, including checking of data stationary

ii. Identification of ARIMA model through autocorrelation function and partial autocorrelation function

iii. Estimation of ARIMA model parameters: p, d, and q

iv. Determination of ARIMA model equations

v. Forecasting.

3.3 Identification

Identification requires calculation and general review of the results of the autocorrelation function (ACF) and the parisal autocorrelation function (PACF). The results of these calculations are needed to determine the appropriate ARIMA model, whether ARIMA $(p, 0, 0)$ or AR (p), ARIMA $(0, 0, q)$ or MA (q), ARIMA $(p, 0, q)$ or ARMA (p, q), ARIMA (p, d, q). Meanwhile, to determine the presence or absence of the d model value, it is determined by the data itself. If the data form is stationary, d is 0, while the data form is not stationary, the value of d is not equal to 0 $(d > 0)$. Likewise, the dual seasonal ARIMA model also refers to the autocorrelation function (ACF) and partial autocorrelation function (PACF) as well as knowledge of the system or process being studied.

Identification can be done after fixed time series data. The application of the model after ACF and PACF data has a tendency according to the reference to **Table 1** and for the seasonal data patterns determined by referring to **Table 2** [11].

3.4 Parameter approximation

There are two basic ways to get this parameter:

a. By trial and error, test several different values and choose one of these values (or a set of values, if more than one parameter is estimated) that minimizes the sum of squared residuals.

b. Iterative approach, choosing an initial estimate and then letting the computer correct the iterative approximation.

ACF patterns	PACF patterns	ARIMA parameters
Heading to zero after lag q	Decreasing gradually/bumpy	ARIMA $(0, d, q)$
Decreasing gradually/bumpy	Heading to zero after lag q	ARIMA $(p, d, 0)$
Decreasing gradually/bumpy (until lag q is still different from zero)	Decreasing gradually/bumpy (until lag q is still different from zero)	ARIMA (p, d, q)

Table 1.
PACF and ACF patterns.

Model	ACF	PACF
AR (p)	Dies down (decreases exponentially) in seasonal lags	Cut off after lag p^s
MA (q)	Cut off after lag q^s	Dies down (decreases exponentially) in seasonal lags
ARMA (p,q)	Dies down (decreases exponentially) in seasonal lags	Dies down (decreases exponentially) in seasonal lags

Table 2.
PACF and ACF seasonal patterns.

3.5 Parameter testing

Parameter testing phase is to test whether the selection of parameters p, d, q is true and correct. The model is said to be good if the error value is random, meaning that it no longer has a certain pattern. In other words, the model obtained can capture well the existing data patterns. To see the error value of the test carried out testing the value of the autocorrelation coefficient of the error, using one of the following two statistics:

1. Q Box dan Pierce Test

$$Q = n' \sum_{k=1}^{m} r_k^2 \quad (5)$$

2. Ljung-Box Test

$$Q = n'(n'+2) \sum_{k=1}^{m} \frac{r_k^2}{(n'-k)} \quad (6)$$

Spread by chi squared (χ^2) with free degrees $(db) = (m - p - q - P - Q)$
Where

$$n' = n - (d + SD) \quad (7)$$

3.6 Testing criteria

If $Q \leq \chi^2(\alpha, db)$, meaning: error value is random (model is accepted)
If $Q > \chi^2(\alpha, db)$, meaning: error value is not random (model cannot be accepted

3.7 Parameter estimation

This study uses the least squares method in estimating parameters [15]. The ARIMA model parameters are based on the time series observed with $Z_1, Z_2, \ldots, Z_1$. The quadratic method assumes that the best curve is the curve that has the least square error of the data set. The parameter values of the ARIMA models $p, d,$ and q are determined through the stationary ACF and PACF chart plots.

3.8 Measuring accuracy level of forecasting result

Basically, to measure the accuracy of forecasting result can be done by various methods. Some statistical methods such as as Root Mean Square Error

(RMSE), Mean Absolute Error (MEA) and Mean Absolute Percentage Error (MAPE). In this research. MAPE is used as a standard measurement of the accuracy offorecasting result. MAPE is defined as follows [13]

$$\mathrm{MAPE} = \frac{\sum_{i=1}^{n} \left| \frac{z - \hat{Z}_i}{Z_i} \right|}{n} \times 100\% \tag{8}$$

Where Z_i and $\hat{Z}_i$ is the actual and predicted values, while n is the number of predicted values.

3.9 Electric load cluster modeling

Cluster analysis performed in this study refers to the statistical description of the analysis technique. Descriptive statistics are methods relating to the collection and presentation of a group of data so as to provide useful information [16]. This description analysis includes several things, namely: frequency distribution, measurement of central tendency, and measurement of variability [17].

The data that has been obtained from a study which is still in the form of random data that can be made into grouped data is data that has been arranged into certain classes. Lists containing grouped data are called frequency distributions or frequency tables. Frequency distribution is the arrangement of data according to certain interval classes or according to certain categories in a list. Frequency distribution can be presented in groups, distribution based on rank order or ranking of distribution classes, distribution in groups, and distribution charts.

Measuring central tendency is a statistical analysis that specifically describes a representative score. The central tendency shows the location of the largest part of the value in the distribution including a general description of data frequencies such as mode, media, and mean or mean count.

While the measurement of variability to describe the degree of dispersion of quantitative data. This measure consists of interquartile range, quartile deviation, mean deviation, standard deviation and coefficient of variation, and variance. Measurement of variability serves to determine the homogeneity or heterogeneity of data. A data may have the same central tendency value but have different variance values.

4. DSARIMA-based load forecasting

The data used in this study is the consumption of electric power every 30 minutes during January 2, 2009 to November 19, 2011 in the Generating Unit service, the National Electricity Company in Gresik City, Indonesia.

The data is distributed on: 1. Data for training during January 2, 2009 to November 12, 2011, 2. Data for testing with the assumption of real data compared to training data from forecasting results during November 13–19, 2011.

Statistical Analysis System (SAS) is used as a simulation of electricity load forecasting and Minitab programming is used to analyze the electricity load cluster model.

4.1 Parameter identification

To identify data, the first step that must be taken is to plot the time series of the data. The time series plot is displayed to see the data patterns and stationarity of the data which aims to determine the ARIMA model. The pattern of data as shown in **Figure 1** is very volatile. This condition is likely influenced by the integrated power distribution system in the Java-Madura-Bali Indonesia interconnection system.

When referring to **Figure 1(a)**, it can be seen that the data are not stationary in variance or mean. For more details, it will be seen in the autocorrelation function as shown in **Figure 2**. And if it refers to time series patterns there is a tendency for the data to contain seasonal patterns as shown in **Figure 1(b)**.

The data is not stationary in the variance, so it is necessary to transform the data as follows. Testing stationarity in variance if the p-value or $\lambda = 1$. Based on the results of the transformation, the data is not stationary in the variance marked with the value $\lambda = -0.13$ as shown in **Figure 3a**. After going through the process of transformation the data becomes significant with the value $\lambda = 1$ as shown in **Figure 3b**.

After the data is transformed it will be transformed back to get the active data value, as follows

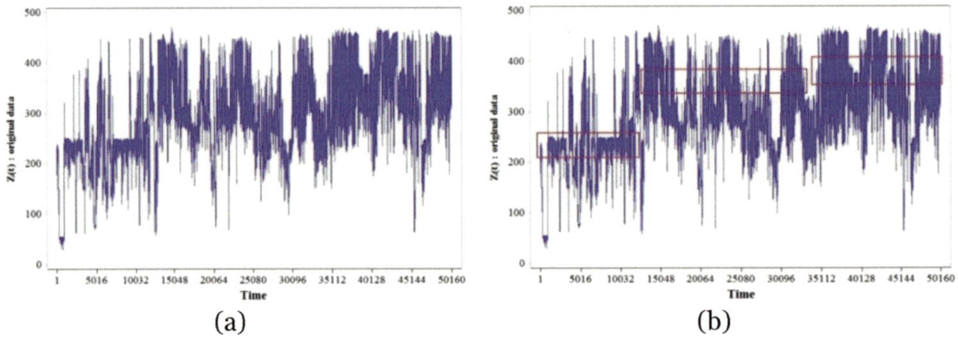

Figure 1.
(a) Data plot of electricity usage every 30 minutes during January 2, 2009 to November 12, 2011; (b) plot of electrical load data with seasonal patterns (red box).

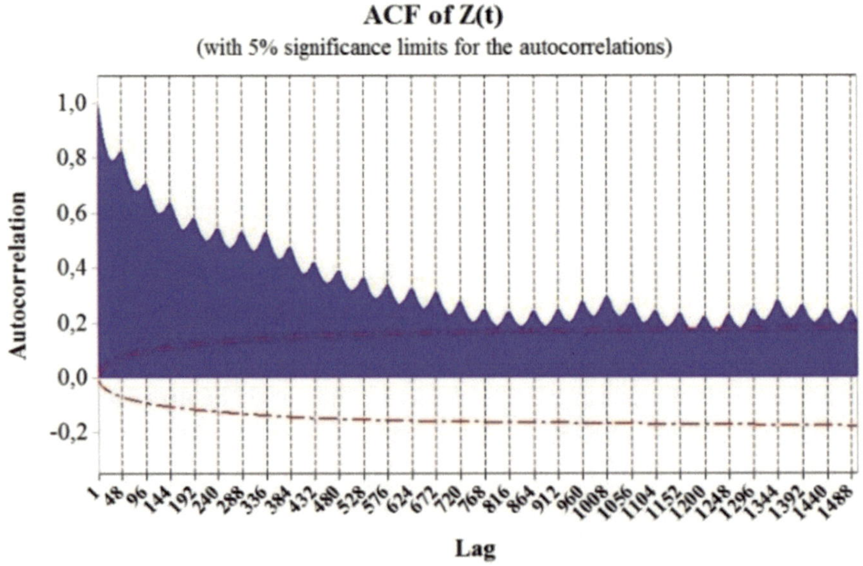

Figure 2.
ACF plot data.

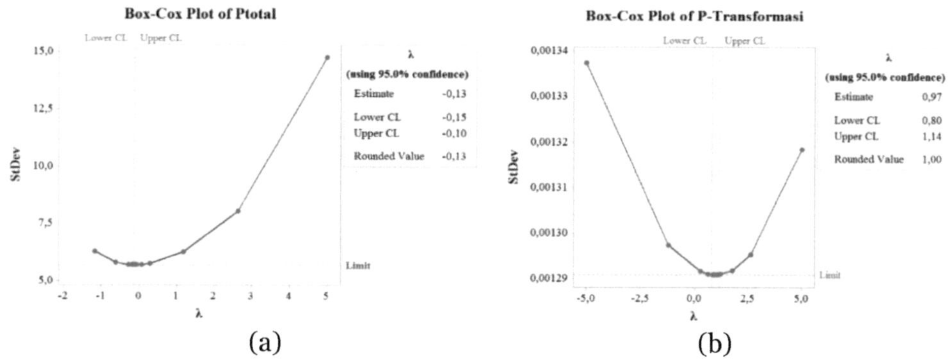

Figure 3.
(a) Box-Cox transformation; (b) after transformation.

$$Z_t^* = Z_t^{-0,13} \tag{9}$$

Then

$$Z_t = \left(Z_t^*\right)^{-\frac{100}{13}} \tag{10}$$

The data is stationary in variance, but the transformation results in **Figure 2b** are not stationary in the mean. Data has not shown a constant value in the middle. The stationarity of the data can also be seen through the plot of the autocorrelation function (ACF). From **Figure 2**, it can be seen that the coefficient of autocorrelation is significantly different from zero and slowly decreases. The pattern shows that the data is not stationary in particular not stationary in the mean, while the ARIMA method requires data that is stationary.

The ACF plot also shows that there are strong indications of having a seasonal pattern in both daily and weekly seasonal averages as shown in **Figure 4**, below.

In **Figure 4a**, it can be seen that the electricity load data has a seasonal pattern that is the daily seasonal as seen in lags 48, 96, 144, etc. And in **Figure 4b**, the data also contains weekly seasonal as seen in lag 336, 672, 1008, 1344, etc.

Because the data is not stationary in the mean, it is necessary to do differencing $(d = 1)$. The ACF plot of differencing data results is shown in **Figure 5** below.

Based on the ACF plot in **Figure 5**, it appears that the nonseasonal data has been stationary. However, seasonal plots are still not stationary with an indication that ACF is still falling slowly in daily seasonal lags, ie lags 48, 96, 144, etc., and weekly seasonal lags, ie lags 336, 672, etc.

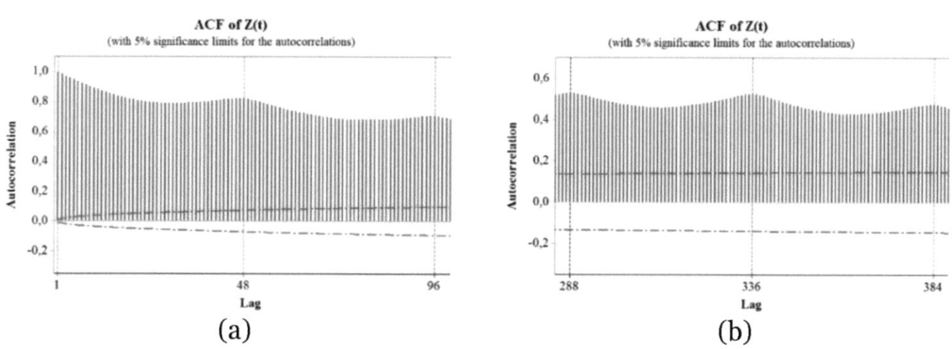

Figure 4.
ACF plots with seasonal patterns: (a) daily seasonal; (b) weekly seasonal.

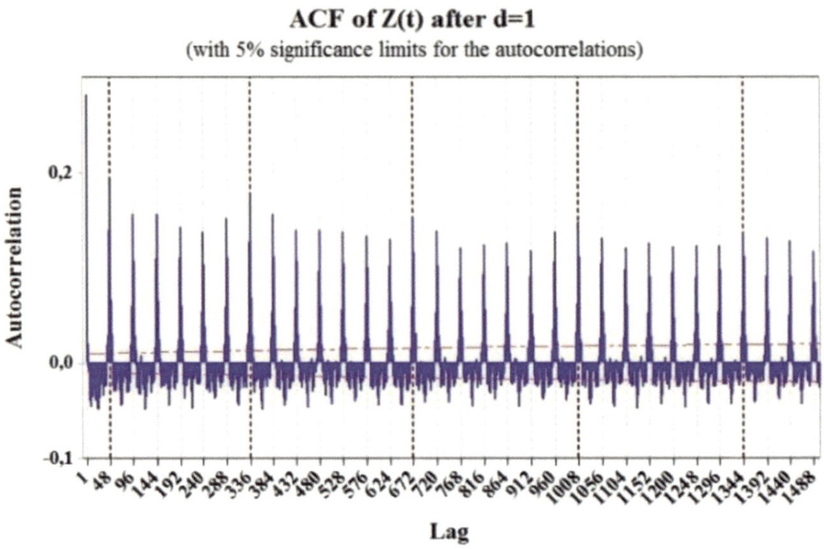

Figure 5.
ACF plot after differencing ($d = 1$.)

It is necessary to do differencing data once more in the seasonal pattern ($d = 1, D = 1, s = 48$). After going through seasonal differencing there are strong indications that the data patterns have been stationary.

Based on the ACF plot for differencing ($d = 1, D = 1, s = 48$) it is clear that the data as a whole has been stationary in the mean. The nonseasonal data plot has been stationary in lags 1, 2, 3, ..., 40. The data pattern tends to dies down and will be cuts off after lag 7 and lag 8 in **Figure 6a**.

The ACF plot for seasonal patterns $s = 48$ after differencing has also been stationary at lags 48, 96, 144, etc. The data pattern tends to be cuts off after lag 48 in **Figure 6b**. The seasonal pattern $s = 336$ tends to be cuts off after lag 336 in **Figure 6c**.

For PACF plots both seasonal ($s = 48$) and ($s = 336$) dies down as shown in **Figure 6d**. Based on the provisions in **Tables 1** and **2**, the parameter identification results can be rewritten in the following **Table 3**.

The ACF and PACF data plots are stationary, the alleged nonseasonal ARIMA models are in accordance with the stationary topology in **Table 1** and the seasonal ARIMA in **Table 2**. The temporary model of ARIMA provisional model is double seasonal based on **Table 3** is DSARIMA $(1, 1, 1)(0, 1, 1)^{48}(0, 0, 1)^{336}$. However, there is a possibility that white noise has not been fulfilled, so it is necessary to add or change the order in accordance with the test.

4.2 Parameter estimation

AR and MA coefficients in the DSARIMA model are estimated using the least squares method. The initial estimate that has been obtained is used as the initial value of the estimation method iteratively. Obtained initial estimates of AR and MA coefficients from the interim model DSARIMA (1, 1, 1) $(0, 1, 1)^{48}$ $(0, 0, 1)^{336}$ as shown in **Table 4** in the following.

Based on **Table 4**, AR and MA parameters have met the criteria for white noise with a p-value greater than the error tolerance value α = 5%, with an alpha significance level of less than 0.0001. However, it is necessary to re-test the residual

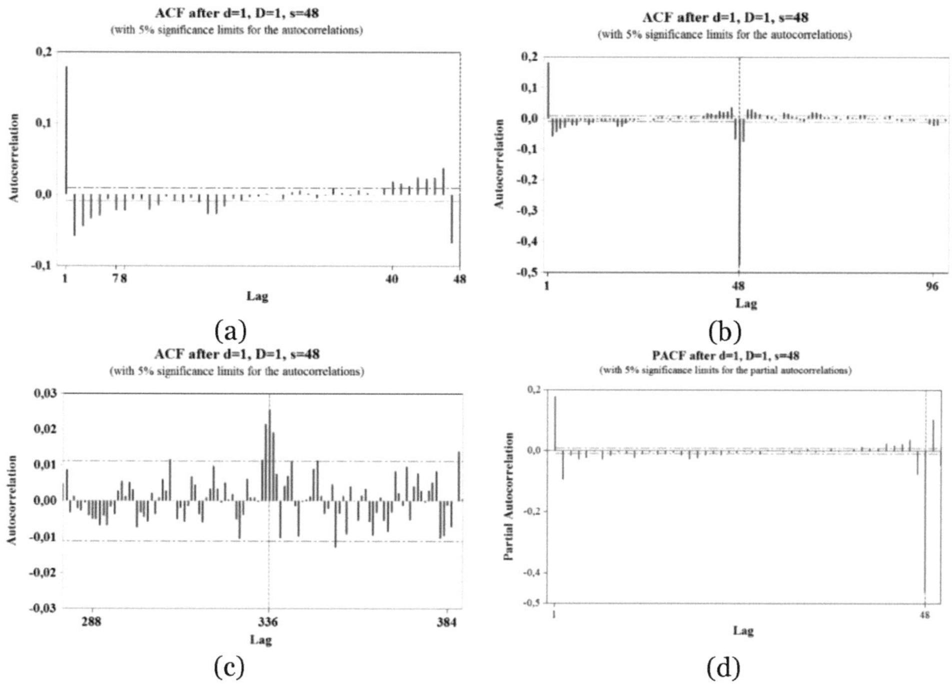

Figure 6.
ACF and PACF plot after differencing ($d = 1, D = 1, s = 48$)

Models	ACF	PACF	Estimated parameters
Nonseasonal	Dies down	Dies down	ARMA (1, 1)
Seasonal ($s = 48$)	Cuts off	Dies down	MA $(1)^{48}$
Seasonal ($s = 336$)	Dies down	Dies down	MA $(1)^{336}$

Table 3.
Identification plots for ACF and PACF.

Parameter	Estimate	Standard error	*t* value	Approx Pr > |*t*|	Lag
MA 1.1	−0.35184	0.01899	−18.53	<0.0001	1
MA 2.1	0>95734	0.0013007	736.02	<0.0001	48
MA 3.1	−0.04526	0.0045103	−10.03	<0.0001	336
AR 1.1	−0.14,578	0.02006	−7.27	<0.0001	1

Table 4.
An output SAS of model with CLS iterative.

assumptions which include the white noise assumption and meet the independent criteria and are normally distributed $(0, \sigma^2)$.

Ljung-Box Test is used to check the assumption of independence from residuals with the following hypotheses:

$$H_0 : \rho_1 = \rho_2 = \dots = \rho_K = 0$$

H_1: there is at least one ρ_i that is not equal to zero for $i = 1, 2, \dots, K$

With an error tolerance of 5%, H_0 is rejected if the ρ-value $< \alpha$, which means the residual does not meet the assumption of white noise. The initial residual tests are shown in **Table 5** below.

Based on the estimated AR and MA coefficient parameters in **Table 5**, the residual normal probability plot must meet the assumption of white noise with a limit of $< \pm \frac{1.96}{\sqrt{n}} \approx \pm 0.009$, where n as many as 50,160 training data. Then based on the initial estimation results in **Table 5**, it is necessary to estimate to meet the white noise assumption, namely by including an estimate on the lag 2, 3, 4, 5, 7, 8, 9, 11, 16, 17, 18, 19, 20, 21, 22, 23, 27, 29, 30, 31, 46, 47, and 48. The results of the residual check are shown in **Table 6** below. The estimation results are significant for seasonal lag, which is lag 48.

Based on residual checking, namely by adding and subtracting AR and MA parameters, it can be seen that all lags have met the assumption of white noise with a limit of $< \pm \frac{1{,}96}{\sqrt{n}} \approx \pm 0{,}009$ (see ACF Results). The best iteration results of the AR and MA parameters are shown in **Table 7** below.

Based on **Table 7**, the DSARIMA model is obtained with the coefficients $([1, 2, 5, 6, 7, 11, 16, 18, 35, 46], 1, [1, 3, 13, 21, 27, 46])\ (1, 1, 1)^{48} (0, 0, 1)^{336}$, which have met the assumption of white noise.

To Lag	ChiSq	DF	Pr > ChiSq	ACF results					
6	153.39	2	<0.0001	−0.002	−0.019	−0.041	−0.017	−0.028	−0.008
12	274.15	8	<0.0001	−0.033	−0.027	−0.0114	−0.009	−0.014	−0.007
18	342.13	14	<0.0001	−0.009	−0.009	−0.008	−0.017	−0.016	−0.023
24	422>74	20	<0.0001	−0.023	−0.018	−0.020	−0.011	−0.013	−0.003
30	43.05	26	<0.0001	−0.009	−0.008	−0.017	−0.008	−0.017	−0.014
36	489.03	32	<0.0001	−0.011	−0.009	−0.002	0.000	−0.010	0.000
42	497.60	38	<0.0001	−0.007	−0.008	0.002	−0.005	−0.004	0.003
48	804.03	44	<0.0001	0.001	0.002	0.006	0.018	0.044	0.060

Table 5.
An output SAS of model with ACF check of residuals.

To Lag	ChiSq	DF	Pr > ChiSq	ACF results					
6	—	0	—	0.000	0.000	−0.002	0.004	0.001	0.002
12	—	0	—	−0.005	−0.002	0.008	0.000	−0.007	−0.004
18	—	0	—	0.005	0.001	−0.008	0.001	−0.003	−0.003
24	18>10	5	0.0028	−0.007	−0.000	0.001	0.003	−0.001	0.002
30	24.78	11	0.0098	−0.005	−0.005	−0.002	0.009	0.0011	−0.001
36	31.03	17	0.0198	−0.005	−0.004	−0.000	−0.001	−0.004	0.008
42	33.77	23	0.0686	−0.000	−0.005	0.004	0.000	0.000	0.004
48	37.61	29	0.1314	0.005	0.006	−0.002	−0.002	0.003	−0.000

Table 6.
An output SAS of model with ACF check of residuals.

Parameter	Estimate	Standard error	*t* value	Approx Pr > |*t*|	Lag
MA 1.1	0.934	0.01770	52.78	<0.0001	1
MA 1.2	−0.077	0.0072138	−10.64	<0.0001	3
MA 1.3	0.008	0.0038171	2.18	0.0293	13
MA 1.4	0.00685	0.0031724	2.16	0.0309	21
MA 1.5	0.017	0.0027856	5.92	<0.0001	27
MA 1.6	0.059	0.0067600	8.67	<0.0001	46
MA 2.1	0.98	0.0009744	1003.38	<0.0001	48
MA 3.1	−0.0364	0.0045572	−7.98	<0.0001	336
AR 1.1	1.1464	0.01855	61.81	<0.0001	1
AR 1.2	−0.295	0.0087427	−33.79	<0.0001	2
AR 1.3	−0.0104	0.0052195	−2.00	0.0454	5
AR 1.4	0.0189	0.0067496	2.80	0.0051	6
AR 1.5	−0.0234	0.0047509	−4.93	<0.0001	7
AR 1.6	−0.004	0.030582	−1.29	0.1958	11
AR 1.7	−0.0083	0.0033299	−2.49	0.0126	16
AR 1.8	−0.0125	0.0033252	−3.77	0.0002	18
AR 1.9	−0.007	0.0022520	−3.26	0.0011	35
AR 1.10	0.07	0.0067089	10.62	<0.0001	46
AR 2.1	0.03	0.0050410	5.86	<0.0001	48

Table 7.
An output SAS of model with CLS iterative.

4.3 Electrical load forecasting results

Based on the final results of the estimated parameters in **Table 4** the ARIMA coefficient parameters are obtained as follows: AR (1.1) = 1.1464, AR (1.2) = − 0.295, AR (1.3) = − 0.0104, AR (1, 4) = 0.0189, AR (1.5) = − 0.0234, AR (1.6) = − 0.004, AR (1.7) = − 0.0083, AR (1.8) = − 0.0125, AR (1.9) = − 0.0074, AR (1.10) = 0.07, AR (2.1) = 0.03, MA (1.1) = 0.934, MA (1.2) = − 0.077, MA (1.3) = 0.008, MA (1.4) = 0.00685, MA (1.5) = 0.017, MA (1.6) = 0.059, MA (2.1) = 0.98, MA (3.1) = − 0.0364.

Based on the prediction model parameters obtained DSARIMA models $([1,2,5,6,7,11,16,18,35,46],1,[1,3,13,21,27,46])(1,1,1)^{48}\ (0,0,1)^{336}$ with the model equation as follows:

$$\begin{gathered}\left(1-1.1464B+0.295B^2+0.0104B^5-0.0189B^6+0.0234B^7+0.004B^{11}\right.\\ \left.+0.0083B^{16}+0.0125B^{18}+0.0074B^{35}-0.07B^{46}\right)\left(1-0.03B^{48}\right)Z_t^*=\\ \left(1-0.934B+0.077B^3-0.008B^{13}-0.00685B^{21}-0.017B^{27}\right.\\ \left.-0.059B^{46}\right)\left(1-0.98B^{48}\right)\left(1+0.0364B^{336}\right)a_t\end{gathered}$$

After going through a reverse transformation Z_t electrical load for the comparison of predicted results with actual data (testing) in **Figure 7** below.

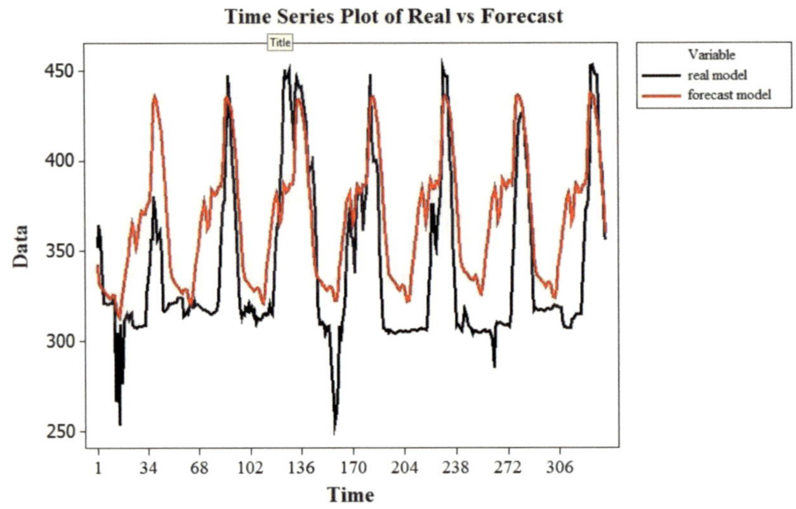

Figure 7.
Comparison of actual power with forecast power.

4.4 Model testing and measuring forecasting accuracy

Accuracy testing between actual power data and prediction results. Test using the MAPE procedure and obtained at 1.56 percent.

5. Electric load modeling

The application of descriptive analytic methods in this book is presented to obtain significant information in managing optimal electrical energy as the author did [18]. Through frequency distribution, data can be arranged based on certain criteria. Data categories are presented based on rank orders that contain ranking data from the top or highest load to the lowest data value.

5.1 Data distribution forecasting results

This electricity load forecasting data is a usage data for a week at intervals every half hour measurement at the power generation. This electricity load forecasting data sample is 336 (N = 336) with mean of 370.56 MWh, meaning that the value is centered at 370.526 MWh. Standard deviation of 36.2582 or the value of this deviation is not too large, this shows the diversity of data is not too large, which means the data is homogeneous.

Descriptive Statistics

N	N*	Mean	StDev	Median	Minimum	Maximum
336	0	370,526	36,2584	370,551	312,912	445,746

Furthermore, forecasting the data shown in the time measurements every half-hour of electric power consumption in the load center in **Figure 8** below.

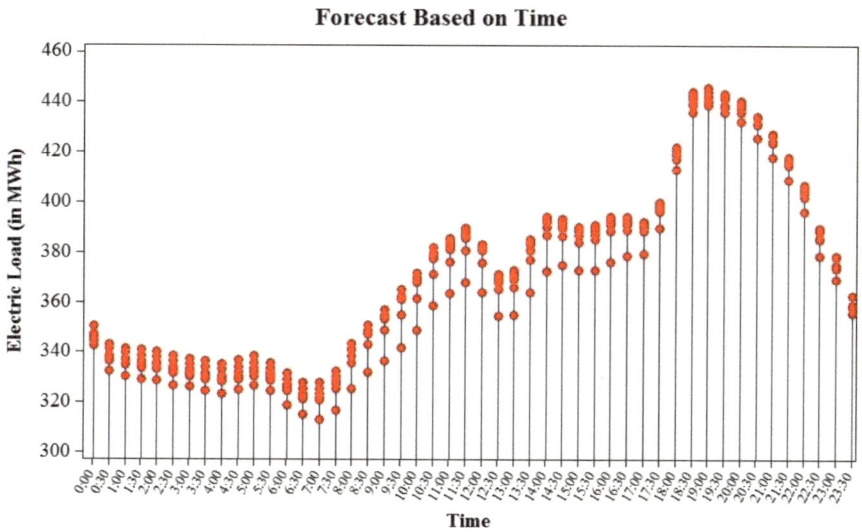

Figure 8.
Plot data forecasting.

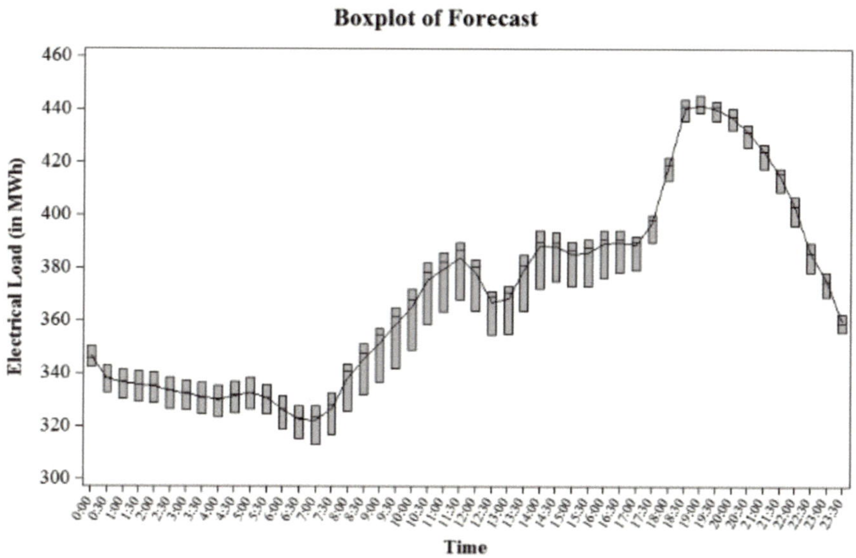

Figure 9.
Graph forecast boxplot.

Visualizations in other forms can be displayed in the form of boxplot graphics. **Figure 9** shows of range (in a box) every hour of measurement and the average value line of every half hour of measurement.

Figure 9 shows that data tend to be at the minimum level, first quartile and the median value. Electricity load increases at third quartile intervals and the maximum load. This condition occurs between 18:30 until 21:30 at night.

Each measurement of electric power absorption at the load center has a peak load. Based on the measurement data, it can be seen that the peak power load absorption occurs at 19:00 and generally the peak load tendency occurs at that hour.

No	Days	Mean	StDev	Median	Minimum	Peak Load	Time of peak load
1	Friday	375.143	35.4253	375.832	327.509	444.234	19:00
2	Saturday	373.635	36.2699	375.208	325.378	445.746	19:00
3	Sunday	361.193	36.8101	357.005	312.912	438.985	19:00
4	Monday	368.672	36.5413	368.417	320.639	440.478	19:00
5	Tuesday	370.616	36.2821	371.793	321.685	439.731	19:30
6	Wednesday	371.619	36.3137	372.718	322.735	441.976	19:00
7	Thursday	372.806	36.6616	374.899	323.262	442.727	19:30

Table 8.
Daily data samples.

Henceforth processing this distribution data through seasonal data that can be presented in the form of daily data, as follows.

The sample data used is Friday data and then the data will be presented in **Table 8** below.

Friday's electricity load data—samples of electric load data are 48 (N = 48) with mean of 375.143 MWh, meaning that the value is centered at 375.143 MWh. Standard deviation of 35.4253 or the value of this deviation is not too large, this shows the diversity of data is not too large, which means the data is homogeneous.

Descriptive Statistics

N	N*	Mean	StDev	Median	Minimum	Maximum
48	0	375,143	35,4253	375,832	327,509	444,234

On Friday shown in **Figure 10**, the peak load occurred at 19:00 amounting to 444.234 MWh with a minimum electric absorption range of 327.509 MWh. On Friday, the data has mean of 375.143 MWh.

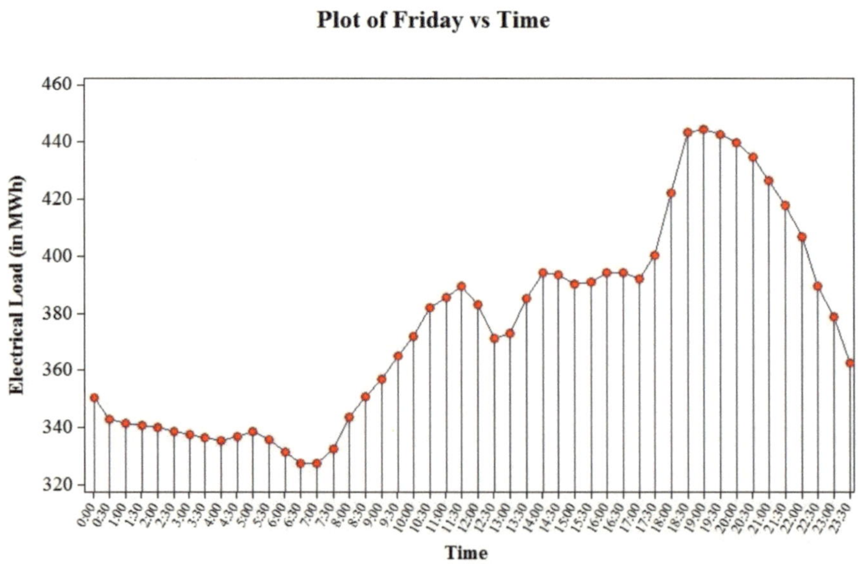

Figure 10.
Data plot on Friday.

Clusters	Interval range	Frequency
1	Min–Q1	97
2	Q1–Median	83
3	Median–Q3	86
4	Q3–Max	70
		N = 336

Table 9.
Range of clusters in the data variant.

Furthermore, seasonal electricity load data on a daily scale can be restated in the form of **Table 8** below.

5.2 Predicted cluster data

In descriptive analysis, frequency distribution, measurement of central tendencies and measurement of variability can be presented in the frequency distribution graph. The purpose of the presentation and information provided in addition to being able to describe the tendency of the data to form certain patterns, this analysis can also be used as a reference for changes in electric power in the power generation system.

The degree of data dispersion can be determined based on the range of interquartile intervals that indicate the homogeneity of the data. In this study, the electrical load cluster is defined as the range of quartile intervals to median value or is shown in the electrical load data below.

Descriptive Statistics: Forecast

Variable	N	Mean	StDev	Variance	Minimum	Q1	Median	Q3	Maximum
Forecast	336	370,53	36,26	1314,67	312,91	335,64	370,55	390,83	445,75
N for									

It can be seen that the data sample with N = 336 has an average of 370.53 MWh which means that the centralized data distribution is rated median. Standard deviation of 36.26 or the value of this deviation is not too large, this shows the diversity of data is not too large, which means the data is homogeneous.

Quartile intervals that divide data over median values form a cluster pattern, with the distribution of data presented in **Table 9** below

An important aspect of this data sample analysis is the presentation of data with seasonal variants. Data development by taking into account the seasonal variants of the hours and daily helped to optimize the management and operational decisions of the generating system both in scheduling and controlling.

6. Conclusion

One of the research trends in electrical engineering is time series analysis. This research includes forecasting studies and modeling of electrical load clusters. The time series analysis method is very suitable with the characteristics of the electrical load that is always fluctuating. This method is also able to produce different data or not included in the training data process.

For the purposes of this electrical load research, forecasting study using the DSARIMA method is an appropriate choice. This method accurately considers the seasonal parameters of the electricity load with MAPE of 1.56 percent when compared with the actual data.

Whereas the modeling of electrical load clusters based on descriptive analytic methods, obtained knowledge of the dynamics of electrical loads. The electrical load pattern has seasonal characteristics at daily and weekly intervals. This pattern forms a unique load characteristic at all times.

So, forecasting studies and modeling of electricity load clusters are able to answer the challenges of electricity energy utilization policies and the operation of generating systems that are able to maintain the balance of supply and demand.

Nomenclature

T	period of time (hours)
$P_{average}$	average load in period T (watts)
P_{peak}	peak load in the T (watts)
$\boldsymbol{p, d, q}$	nonseasonal parts of the model
P, D, Q	seasonal parts of the model
S_1, S_2	1st and 2nd period seasonal
D_1, D_2, d	order of differences
S	number of period per season
m	maximum lag time
r_k	autocorrelation or time-lag , 2, 3, … , k
Z_t	time series process in period T
Z_t^*	forecasting process in transformation in period T
Q_1	quartile 1
Q_3	quartile 3

Greek symbols

λ	Box-Cox transformation number
α_t	white noise
$\theta_q(B)$	regular MA polynomials of order q
$\Theta_{Q_1}(B^{S_1}), \Theta_{Q_2}(B^{S_2})$	MA polynomials of orders
$\varphi_p(B)$	regular AR polynomials of orders p
$\Phi_{P_1}(B^{S_1}), \Phi_{P_2}(B^{S_2})$	AR polynomials of orders

Abbreviations

MAPE	mean absolute percentage error
MWh	mega watt hours

Author details

Ismit Mado
University Borneo Tarakan, Tarakan City, Indonesia

*Address all correspondence to: ismitmado@borneo.ac.id

References

[1] Tsekouras GJ, Dialynas EN, Hatziargyriou ND, Kavatza S. A on-linier multivariable regression model for midterm energy forecasting of power systems. Electric Power Systems Research. 2007;**77**(12):1560-1568

[2] McSharry PE, Bouwman S, Bloemhof G. Probabilistic forecasts of the magnitude and timing of peak electricity demand. IEEE Transactions on Power Systems. 2005;**20**(2):1166-1172

[3] Gonzalez-Romera E, Jaramillo-Moran MA, Carmona-Fernandez D. Monthly electric energy demand forecasting based on trend extraction. IEEE Transactions on Power Systems. 2006;**21**(4):1946-1953

[4] Taylor JW, Mc Sharry PE. Short-term load forecasting methods: An evaluation based on european data. IEEE Transactions on Power Systems. 2007;**22**(4):2213-2219

[5] Hahn H, Meyer-Nieberg S, Pickl S. Electric load forecasting methods: Tool for decision making. European Journal of Operational Research. 2009;**199**(3): 902-907

[6] Nia MM, Din J, Lam HY, Panagopoulos AD. Stochastic approach to a rain attenuation time series synthesizer for heavy rain regions. International Journal of Electrical and Computer Engineering. 2016;**6**(5):2379

[7] Kamley S, Jaloree S, Thakur RS. Performance forecasting of share market using machine learning techniques: A review. International Journal of Electrical and Computer Engineering. 2016;**6**(6):2088-8708

[8] Mado I, Soeprijanto A, Suhartono S. Applying of double seasonal ARIMA model for electrical power demand forecasting at PT. PLN Gresik Indonesia. International Journal of Electrical and Computer Engineering. 2018;**8**(6):4892-4901. ISSN: 2088-8708. DOI: 10.11591/ijece.v8i6

[9] Suswanto D. Karakteristik beban tenaga listrik. In: Suswanto D, editor. Sistem Distribusi Tenaga Listrik. Indonesia: Departement of Electrical Engineering, Universitas Negeri Padang; 2009

[10] Makridakis S, Hibon M. ARMA models and the Box-Jenkins methodology. Journal of Forecasting. 1997;**16**(3):147-163

[11] Wei WW. Time Series Analysis: Univariante and Multivariante Methods. Vol. 2. Boston, MA: Pearson Addison Wesley; 2006;1:1:108-109

[12] Soares LJ, Medeiros MC. Modeling and forecasting short-term electricity load: A comparison of methods with an application to Brazilian data. International Journal of Forecasting. 2008;**24**(4):630-644

[13] Mohamed N, Ahmad MH, Ismail Z, Suhartono S. Short term load forecasting using double seasonal ARIMA model. In: Proceedings of the Regional Conference on Statistical Sciences. Vol. 10. 2010. pp. 57-73

[14] Kim SY, Jung HW, Park JD, Baek SM, Kim WS, Chon KH, et al. Weekly maximum electric load forecasting for 104 weeks by seasonal ARIMA model. Journal of the Korean Institute of Illuminating and Electrical Installation Engineers. 2014;**28**(1):50-56

[15] Cryer JD, Chan KS. Time Series Analysis: With Application. R. Springer Science & Business Media; 2008;7:154-158

[16] Walpole RE. Pengantar Statistika. Jakarta: PT Gramedia Pustaka Utama; 1993. ISBN 979-403-313-8

[17] Wiyono BB. Statistik Pendidikan, Buku Ajar Mata Kuliah Statistik. 2001

[18] Mado I, Soeprijanto A, Suhartono S. Electrical load adat clustering in PJB UP Gresik based on time series analysis approach. In: Proceedings of the International Conference on Vocational Education and Electrical Engineering. Surabaya, Indonesia; 2015. pp. 261-266. Available from: http://digilib.unimed.ac.id/23841/1/Fulltext.pdf

Chapter 7

Seeking Accuracy in Forecasting Demand and Selling Prices: Comparison of Various Methods

Zineb Aman, Latifa Ezzine, Yassine Erraoui, Younes Fakhradine El Bahi and Haj El Moussami

Abstract

The need for a good forecast estimate is imperative for managing flows in a supply chain. For this, it is necessary to make forecasts and integrate them into the flow control models, in particular in contexts where demand is very variable. However, forecasts are never reliable, hence the need to give a measure of the quality of these forecasts, by giving a measure of the forecast uncertainty linked to the estimate made. Different forecasting models have been developed in the past, particularly in the statistical area. Before going to our application on real industrial cases which highlights a prospective study of demand forecasting and a comparative study of sales price forecasts, we begin, in the first section of this chapter, by presenting the forecasting models, as well as their validation and monitoring.

Keywords: forecasts, accuracy, quality of forecasts, demand forecasting, selling price forecasting

1. Introduction

For most companies, forecasting is a prerequisite for effective supply chain management. As explained by Lai et al. [1], forecasting is the basis of all production management systems. The entire supply chain is based on the data from forecast models.

In Ref. [2], the authors show the usefulness of forecasting and planning as a decision-making tool for organizing the supply chain across all horizons of time and at all levels.

In the academic field, forecasting occupies an important place. Given the primordial role of forecasting, we understand why many models have been developed since the beginning of the twentieth century. Research mainly developed from the 1950s onward with the use of mathematical models. A review of the literature was carried out by Stadtler [3]. We find there the interest of forecasting for the global supply chain in order to integrate the different organizations and coordinate their flows in order to satisfy the end consumer.

The various sources for making these forecasts are located throughout the supply chain, including the commercial part of the business. It is the analysis of this source that will help build the basis for future forecasting. In the end, the sources used to build the forecasts are therefore multiple.

2. Application to prospective approach: modeling and forecasting demand using the ARIMA models

In the manufacturing sector, forecasting demand is one of the most crucial problems in inventory management [4]; it can be used in various operational planning activities during the production process: capacity planning and management of used product acquisitions [5].

For both types of push/pull supply chain processes, demand forecasting forms the basis of all CS planning. The "pull" processes in the SC are performed in response to the client's request, while all the "push" processes are performed in anticipation of the client's request [6]. A business needs to know many factors related to forecasting demand. Some of these factors are listed below:

- past requests;
- product delivery time;
- planned advertising or marketing efforts;
- state of the economy;
- price reduction planned; and
- actions undertaken by competitors.

Businesses need to understand these factors before they can choose an appropriate forecasting method as it can be difficult to decide which method is the most suitable for forecasting. Forecasting methods are classified into the following types: time series, causal, qualitative, and simulation [6].

A time series is considered to be a set of observations cited in chronological order [7]. To forecast demand, time series forecasting models are based on historical data. These mathematical models used are based on the assumption that the future is an expansion of the past [8].

Numerous studies on demand forecasting by time series analysis have been carried out in several fields. They include demand forecasts for food sales [9], tourism [10], spare parts [4, 11], electricity [12, 13], automobiles [14], and some other goods and services [15–17].

In this section, we forecast the demand for a product in a food manufacturing operation based on real data, as well as the precision and characteristics of these forecasts.

Our study will be carried out according to the three stages of the Box-Jenkins approach: identification, estimation, and verification. We present the model relating to product demand from January 2010 to December 2015 as shown in **Figure 1**.

2.1 Identification of model

This refers to the initial preprocessing of the data to make it stationary and to the choice of p and q values that can be adjusted during model fitting.

We present the ACF and PACF diagrams of the series in **Figures 2** and **3**, respectively. We find that this series oscillates, respectively, around an average value, and its autocorrelation function decreases to zero point rapidly, which proves the stationarity of the time series studied.

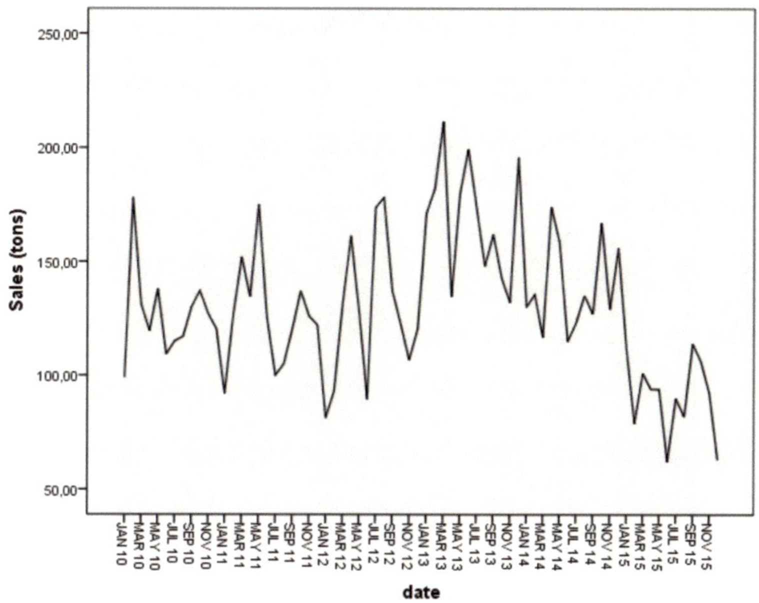

Figure 1.
Evolution of the final product's sales.

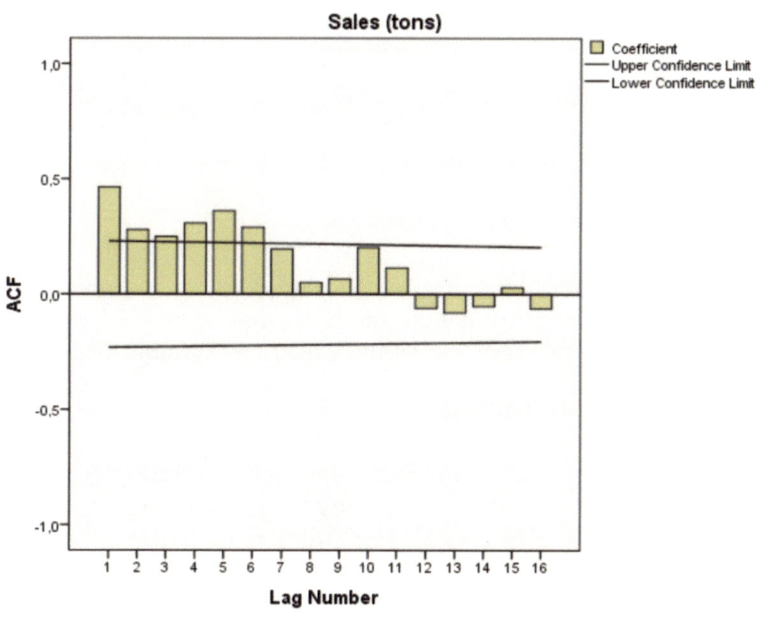

Figure 2.
ACF correlograms of the demand series.

Moreover, to assess whether the data come from a stationary process, we can perform the unit root test: Dickey-Fuller test for stationarity. After carrying out the test on the Xlstat software, the results are grouped in **Table 1**.

H_0: The series has a unit root.

H_1: The series does not have a unit root. The series is stationary.

The null hypothesis H_0 cannot be rejected since the calculated p value is greater than the significance level α set at 0.05. We calculated the risk of rejecting the null hypothesis H_0, while it is true. The risk is 84.38%.

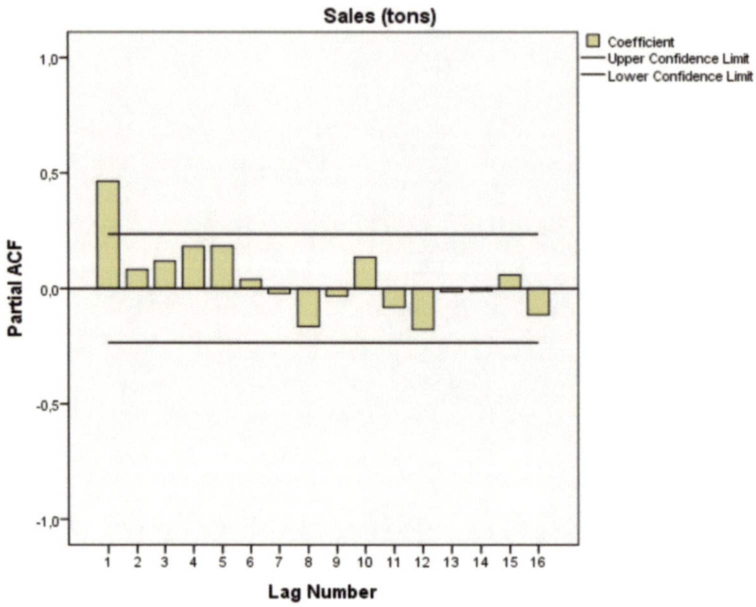

Figure 3.
PACF correlograms of the demand series.

Tau (observed value)	-1.350
Tau (critical value)	-0.717
p (unilateral)	0.844
α	0.05

Table 1.
Test results.

In our study, we checked the stationarity of the series, and we noted from the ACF and PACF correlograms that our model cannot be pure RA or pure MA. Therefore, we tested several models to identify the most suitable for our series.

2.2 Estimation of model coefficients

Using the ARIMA procedure of the SPSS time series module [18], we can estimate the coefficients of our model by providing the parameters p, q, and d [19–22].

The best model is as simple as possible and minimizes certain criteria, namely AIC criteria (Akaike criterion), SBC (Bayesian criterion of Schwarz), variance, and maximum likelihood [23–25]. The chosen model is that of ARIMA (0, 1, 1). For other models, either the Student "T-RATIO" test values are found in the range of ±1.96, or one of the values of the minimization criteria is higher than that found for the ARIMA model (1, 0, 1) with the constant value.

Table 2 presents the values of the different models. From this table, we choose the appropriate model on which we will base ourselves to make our forecasts.

It is clear from **Table 2** that the ARIMA model (1,0,1) is selected because all the coefficients are significantly different from 0 according to the Student test (|T-RATIO|) ≥ 1.96) with an acceptable level of adjustment.

The model residue is stationary and follows a white noise process in the range of ±40. The residue histogram shows whether the distribution of residues approximates a normal distribution. In our case, we have residues that distribute relatively normal around zero and with a relatively low dispersion at a 5% risk.

Characteristics		Models					
		ARIMA (1,0,2)	**ARIMA (2,0,2)**	**ARIMA (1,0,1)**	**ARIMA (1,0,0)**	**ARIMA (0,0,1)**	**ARIMA (1,0,1) without constant**
AR (1)	α_1	0.92913	0.71371	0.90792	0.49434	−0.41704	0.99755623
	SEB	0.104616	0.761758	0.094852	0.1074471	0.1119989	0.00444769
	T value	8.8813204	0.9369292	9.571955	4.600820	−3.723567	224.28617
	p value	0.00000000	0.35216008	0.000000	0.00001823	0.00039384	0.0000000
MA (1)	θ_1	0.52269	0.31779	0.63880			0.71392452
	SEB	0.167073	0.741595	0.161531			0.08579173
	T value	3.1284995	0.4285186	3,954655			8.32160
	p value	0.00258711	0.66964815	0.00018319			0.0000000
AR (2)	α_2		0.19759				
	SEB		0.659442				
	T value		0.2996279				
	p value		0.76538859				
MA (2)	θ_2	0.17062	0.30202				
	SEB	0.142258	0.409353				
	T value	1.1993429	0.7377864				
	p value	0.23455708	0.46322050				
Constant	Cte	124.42969	124.52640	125.53260	128.53887	129.22650	
	SEB	12.608189	12.601296	11.785537	6.5715235	4.8971088	
	T value	9.8689581	9.8820312	10.651411	19.559981	26.388326	
	p value	0.00000000	0.00000000	0.00000000	0.00000000	0.00000000	
AIC		688.86593	690.82312	688.77347	689.37103	693.59055	692.04831

Characteristics	Models					
	ARIMA (1,0,2)	ARIMA (2,0,2)	ARIMA (1,0,1)	ARIMA (1,0,0)	ARIMA (0,0,1)	ARIMA (1,0,1) without constant
SBC	697.9726	702.20645	695.60347	693.92437	698.14388	696.60164
Log likelihood	-340.43297	-340.41156	-341.38674	-342.68552	-344.79527	-344.02415
Error	28.034898	28.221721	28.214812	28.576249	29.443759	28.461048

Table 2.
Coefficients of different models.

The chosen model parameters are presented in **Table 3**.

The developed model is given by Eq. 1.

$$y_t = \delta + \alpha_1 y_{t-1} - \theta_1 \varepsilon_{t-1} + \varepsilon_t \quad (1)$$

With:

- y_t, y_{t-1}: sales of periods t and t–1, respectively.
- $\varepsilon_t, \varepsilon_{t-1}$: residuals of periods t and t–1, and constitute a white nose.
- α_1, θ_1: coefficients of autoregressive and moving average processes, respectively.

We can easily extract from **Table 3** the coefficients of the autoregressive processes and moving averages and inject them into Eq. (1), which becomes:

$$y_t = 125{,}524 + 0{,}90792\, y_{t-1} - 0.6388\, \varepsilon_{t-1} \quad (2)$$

2.3 Accuracy of ARIMA (1, 0, 1) model

In order to assess the accuracy of the developed model, we compare the experimental and simulated sales during the same period. This comparison is drawn up in **Table 4** and reveals that the model selected has great precision and an ability to simulate dynamic sales behavior. Therefore, this model can be used to analyze and model the demand in this food manufacturing.

Figure 4 shows that the model is validated since the predicted demand fluctuates around the adjustment and the forecast demand, which remained between the upper limit and the lower limit.

The error varies, but it is within the tolerance range. In order to minimize this error, we are opting for other approaches in our future work.

2.4 Forecast

Once the appropriate model is defined and validated, we must do the forecasting, using the IBM SPSS forecasting. **Table 4** and **Figure 5** present the results of the

AR (1)	α_1	0.90792
	SEB	0.094852
	T value	9.571955
	p value	0.00000000
MA (1)	θ_1	0.63880
	SEB	0.161531
	T value	3.954655
	p value	0.00018319
Constant	Δ	125.53260

Table 3.
ARIMA model parameters.

Model		73	74	75	76	77	78	79	80	81	82
Sales-Model_1	Prévision	95.12	97.92	100.46	102.77	104.86	106.77	108.49	110.06	111.49	112.78
	UCL	151.41	156.21	160.35	163.95	167.08	169.83	172.25	174.38	176.26	177.93
	LCL	38.83	39.63	40.57	41.59	42.64	43.70	44.74	45.75	46.71	47.63

Table 4.
Forecast sales from January 2016 to October 2016.

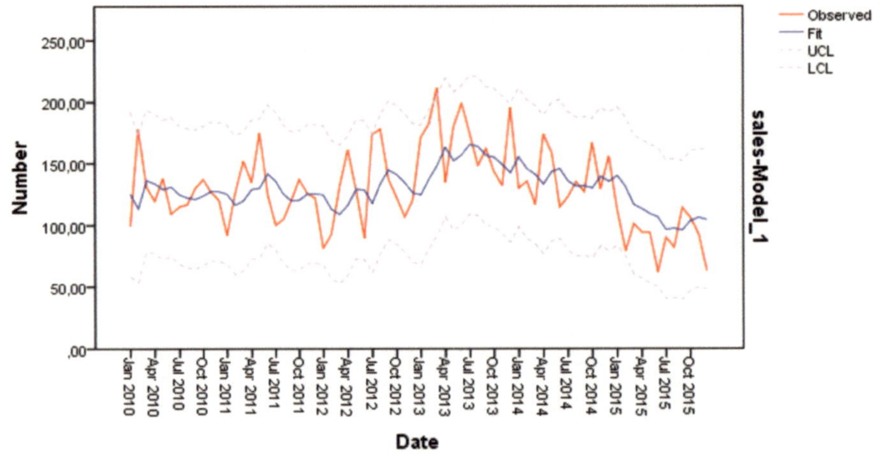

Figure 4.
Sales, fit, LCL, and UCL.

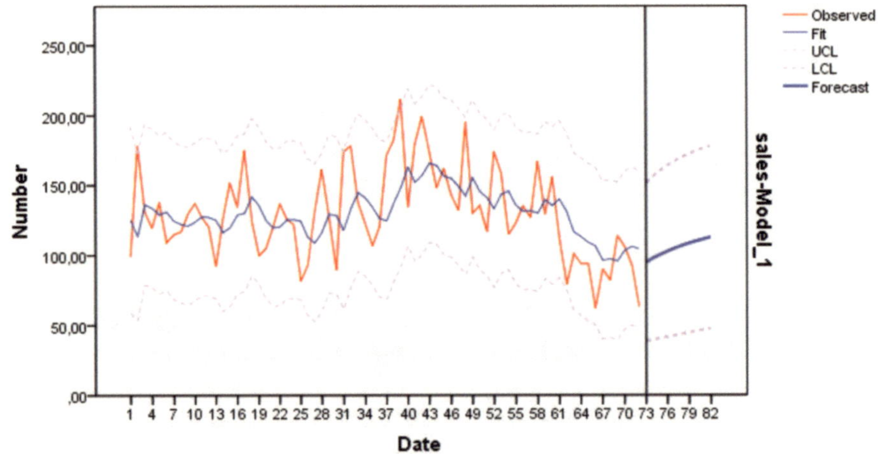

Figure 5.
Sales, fit, LCL, UCL, and forecasting.

sales forecasts that we obtained by applying our ARIMA model (1, 0, 1) for the next 10 months from January 2016 to October 2016.

The chosen model can therefore be used to model and forecast future demand in this food manufacturing. However, each time we have to feed historical data with new data to enrich it and thus improve the new model and forecasts.

The accrue forecasts presented facilitated the production decision in this business. Indeed, the model allowed us to forecast demand and make precise forecasts. Once we have a forecast of demand, it will be much easier to clearly plan the production and thus eliminate the heavy cost losses.

3. Application to comparative approach: comparison of the quality of forecasts obtained in the context of forecasting selling prices

Our second industrial application is devoted to a modeling study and comparative forecast of sales prices using ARIMA models, artificial neural networks, and support vector machines.

In this section, we will model the actual fuel price data named "SSP" in order to make important predictions to determine future selling prices. The model shown in **Figure 6** is based on the price of "SSP" fuel in a petroleum production from January 2012 to December 2016.

3.1 Forecasting using ARIMA models

3.1.1 Determination of the differentiation parameter

Under SPSS, we have drawn the autocorrelation function (ACF) and the partial autocorrelation function (PACF), the results found are presented in **Figures** 7 and 8.

The series has a large number of positive shifts for the autocorrelation function, so it must be differentiated.

The next step is to differentiate the series. You have to differentiate it enough to make it immobile but not drag with an excessive differentiation, which will cause a loss of information and therefore unstable models. In our case, we just had to take $d = 1$ because of the linearity of the trend.

Besides, to decide if the data come from a stationary process or not, we can carry out the unit root test: Dickey-Fuller test for stationarity. After performing the test on the Xlstat software, we grouped the results in **Table 5**.

H_0: The series has a unit root.

H_1: The series does not have a unit root. The series is stationary.

The null hypothesis H_0 must be rejected, and the alternative hypothesis H_1 must be accepted since the calculated p value is less than the significance level α set at 0.05. We calculated the risk of rejecting the null hypothesis H_0, while it is true. The risk is less than 0.92%.

We conclude that our model will have an order of differentiation $d = 1$. We also note that the T-RATIO for the constant of model μ is less than 2 in absolute value. We must therefore deduct it from the model before determining the parameters p and q.

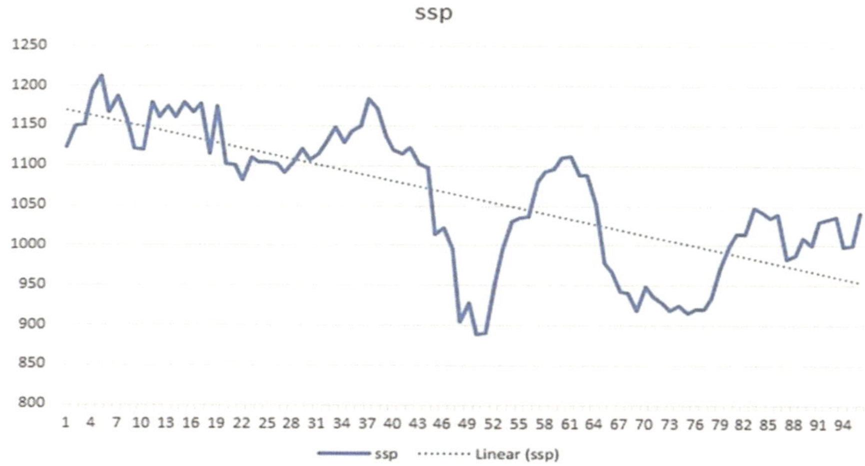

Figure 6.
Selling price of "SSP."

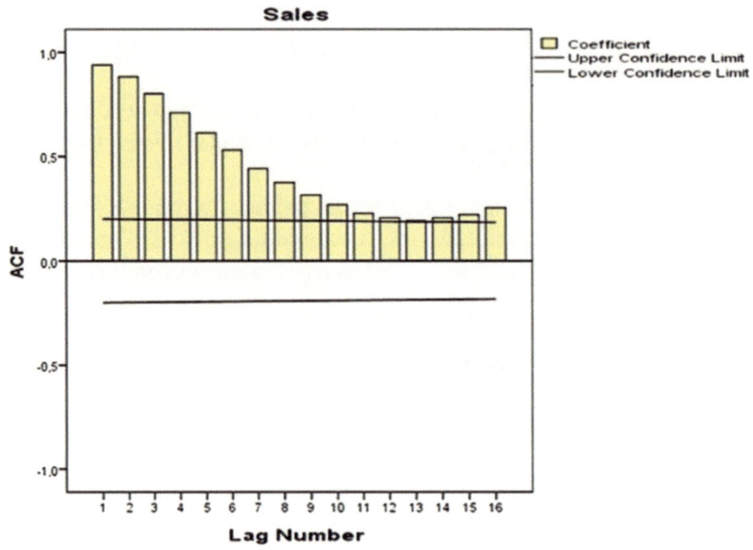

Figure 7.
ACF correlogram for the sales price series.

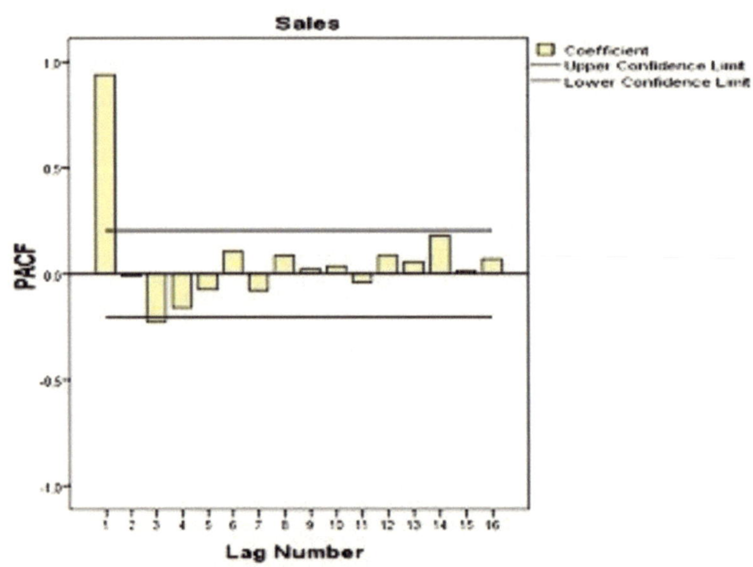

Figure 8.
PACF correlogram for the sales price series.

3.1.2 Determination of the autoregressive parameter

Figures 9 and **10** show the residue curve and the ACF and PACF diagrams of the residues of the ARIMA model (0, 1, 0), respectively.

We can clearly see from **Figures 9** and **10** that the partial autocorrelation has a significant peak at offset 2, and we can then deduce that the differentiated series comprises an autoregressive signature. The parameter p is therefore equal to 1.

However, the T-RATIO for the autoregressive parameter φ_1 is lower in absolute value than 2. So, we cannot retain this model. Similarly, the ARIMA model (2, 1, 0) presents the autoregressive parameters whose T-RATIO is less than 2 in absolute value.

Tau (observed value)	-4.0325
Tau (critical value)	-0.7648
***p* (unilateral)**	0.0092
α	0.05

Table 5.
Test results.

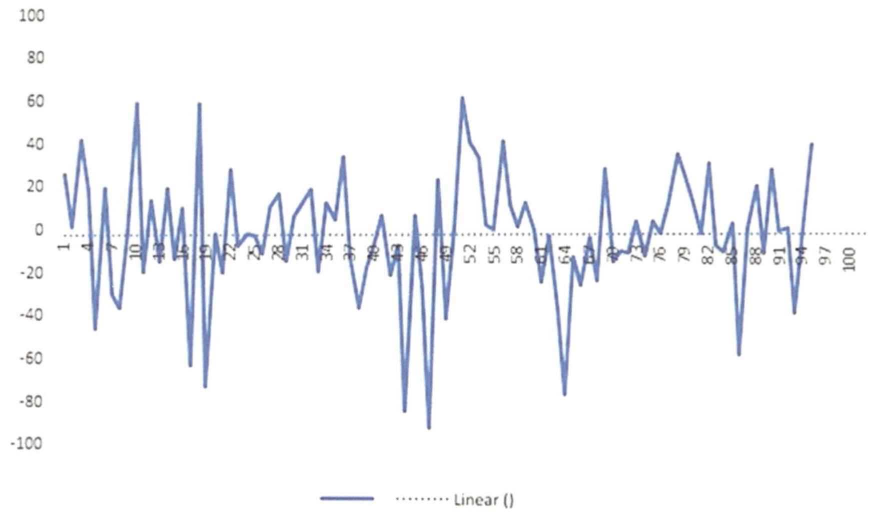

Figure 9.
Residue curve.

3.1.3 Determination of the moving average parameter

Now, the T-RATIO for the moving average parameter θ_1 is lower in absolute value than 2. So we cannot retain this model. Similarly, the ARIMA model (0,1,2) presents moving average parameters whose T-RATIO is less than 2 in absolute value.

3.1.4 Mixed ARIMA model

After several iterations and tests, we concluded that only the ARIMA model (1,1,1) had higher T-RATIOS in absolute value than 2. This is the model we should use to make forecasts.

With the coefficients obtained now, we can write the equation of the model retained as follows:

$$y_t = y_{t-1} - 0.928\left(y_{t-1} - y_{t-2}\right) + 0.873\varepsilon_{t-1} + \varepsilon_t \quad (3)$$

Table 6 lists the forecasts obtained for the first quarter of 2017.

The graph in **Figure 11** proves the adequacy of the ARIMA model (1,1,1) developed, which is very close to the real model.

Table 2 allows us to admit that the chosen model can be used to model and forecast future sales in this petroleum production.

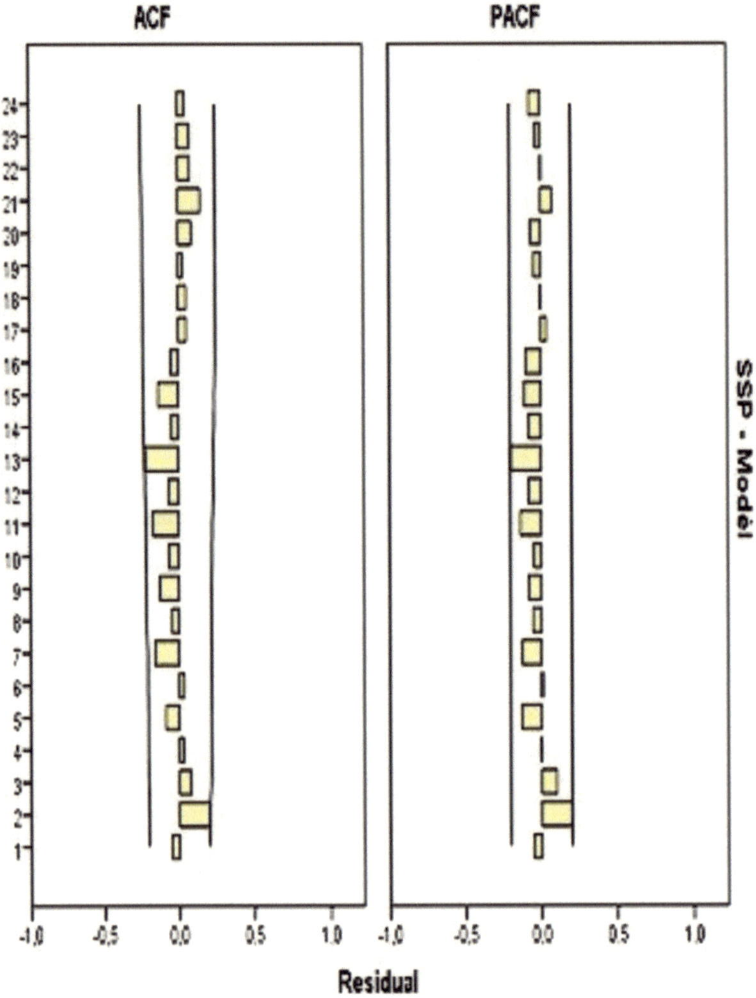

Figure 10.
ACF and PACF diagrams of the residues of the ARIMA model (0,1,0).

Fortnight	Real price	Model	% error
1Q January	1072	1042.49	-2.752798507
2Q January	1074	1043.05	-2.881750466
1Q February	1072	1043.59	-2.650186567
2Q February	1082	1044.21	-3.492606285
1Q March	1084	1044.81	-3.615313653
2Q March	1064	1045.48	-1.740601504

Table 6.
Forecast results for the ARIMA model (1,1,1) [26].

3.1.5 Forecasting using artificial neural networks

The goal here is to develop a relationship between experimental data collected from authentic sources to estimate the selling prices of fuel. We are trying to apply RBF radial-based neural networks, which are based on machine learning approaches

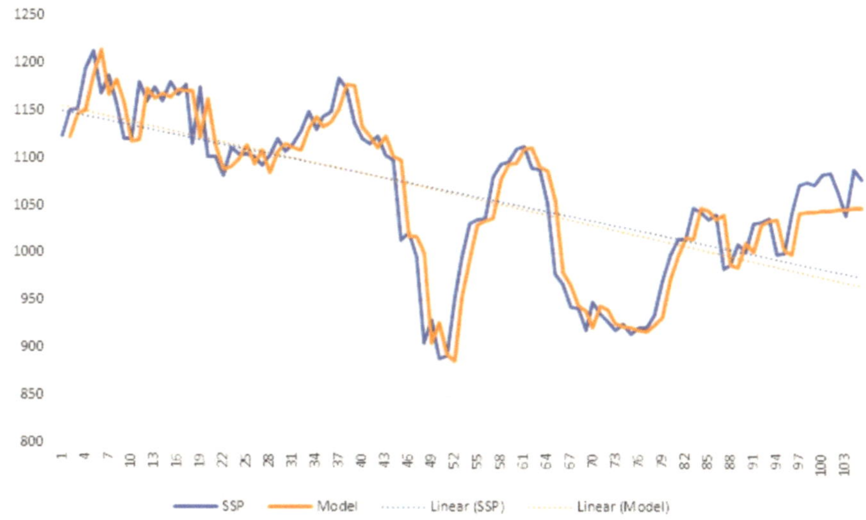

Figure 11.
Results of the ARIMA model (1,1,1) [26].

due to the complex relationships between the input parameter and the output parameter. In this section, we present the modeling approach using this technique to precisely compare it with the ARIMA model used in the previous section.

3.1.6 Model development

The radial basis ANN model (comprising two layers) is trained for implementing the back propagation algorithm to minimize the mean squared error with one parameter (time) as the input and the desired output (fuel selling price). As presented on the visualization of the network shown in **Figure 12**, the first layer has radial basis transfer functions with the maximum number of 80 neurons, and the second layer has a linear transfer function, in order to build a consistent model for providing accurate forecasts [27].

Feature selection is one of the core concepts in machine learning, which hugely impacts the performance of our model. Irrelevant or partially relevant features can negatively impact model performance. Feature selection and data cleaning should be the first and most important step of our model designing. However, in our case, this step may be omitted as long as our point cloud is significant. Subsequently, the dataset was randomly divided into two disjoint subsets of training set (60% of total dataset), which help us train our dataset to find the adequate model and testing set (40% of total dataset) to validate the model found. The training set is applied in order to develop the network. After the training phase, the reliability and accuracy

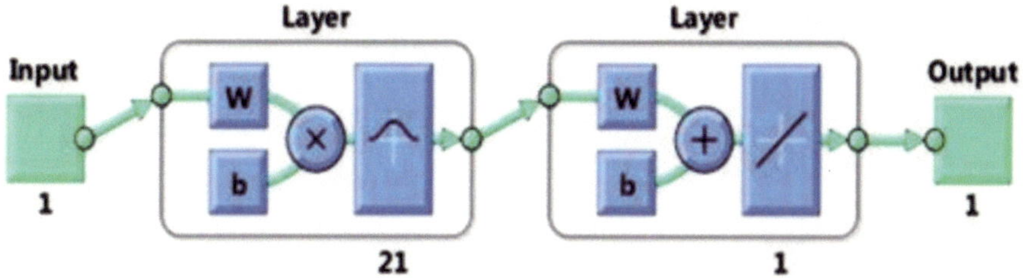

Figure 12.
Visualization of the RBF network.

of the network were perused with the test data. Besides, in our study, we implemented radial basis network of the MATLAB toolbox (i.e., "nwrb"). Furthermore, the Gaussian function is the main kernel function implemented here with the width parameter of 1 [27].

After executing the learning phase, we obtain **Figures 13** and **14** that represent the learning of our database. **Figure 15** represents the error in the training phase. During the test phase, we gave values to the input variable to visualize the results of the output and thus simulate our model.

3.1.7 Error optimization

Optimizing the error consists of a compromise to be made between the various parameters of the network, namely the speed, the objective, the number of neurons, and the number of neurons to be added to the hidden layer. This compromise is made on the basis of several tests of the different combinations carried out. Some of these combinations are presented in **Table** 7.

After making different combinations, we find that the error is considerable for all the compromises. Consequently, no model can adapt to the time series, especially in the long term. The reason behind this result is not only the large fluctuations in

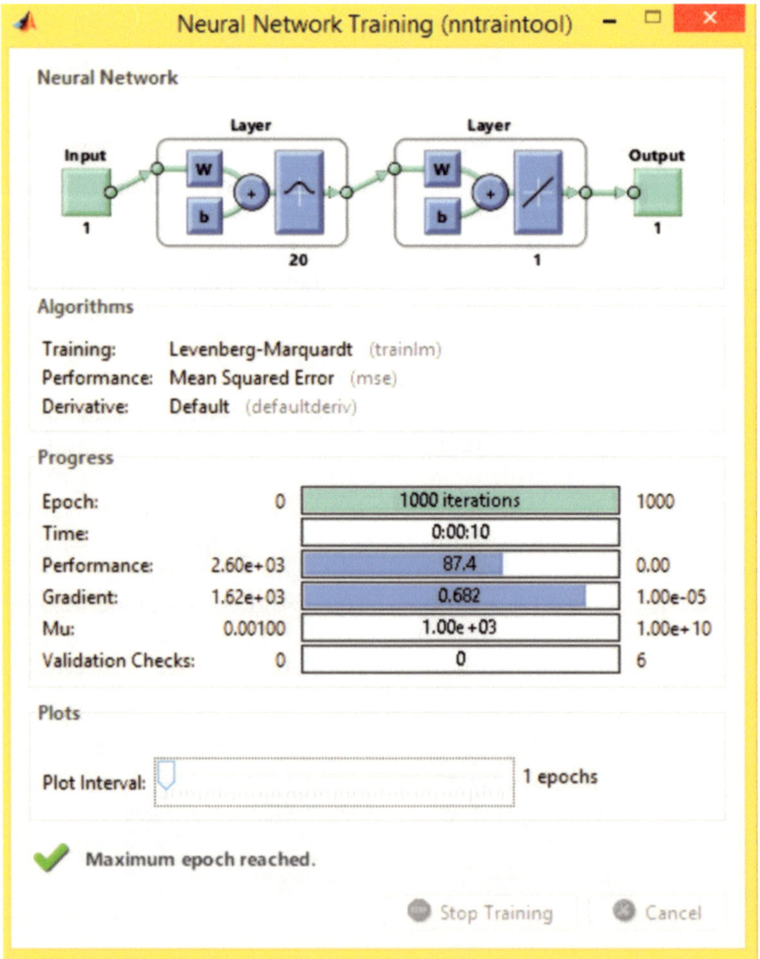

Figure 13.
Training of the RBF network.

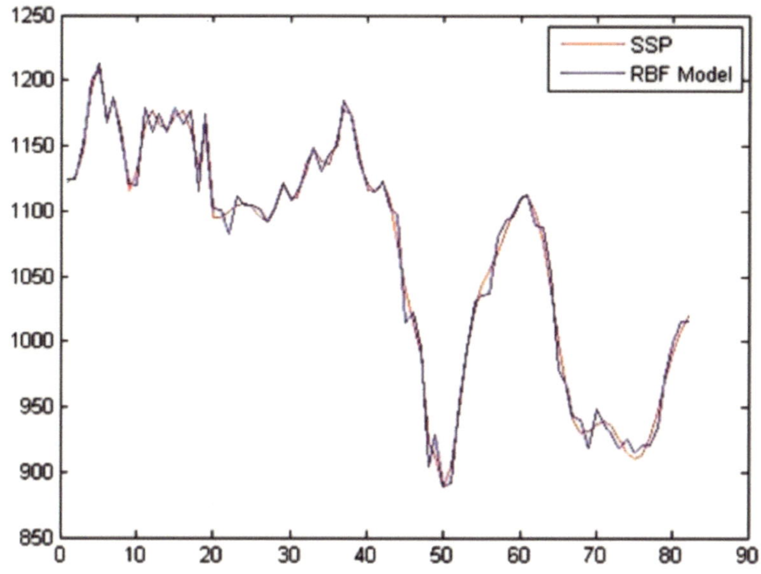

Figure 14.
Training graph of the RBF network.

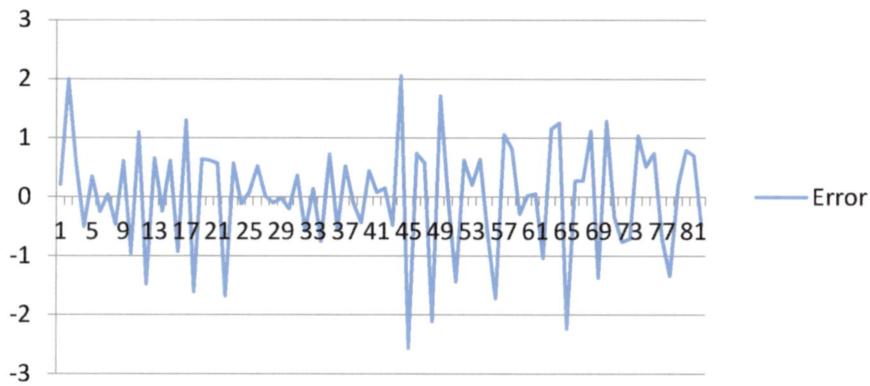

Figure 15.
The graph of error.

Parameters			
Goal	**Spread**	**MN**	**DF**
0 .01	1.5	25	25
0.01	1	25	30
0.01	2	30	30
0.01	0.8	12	30
0.01	1.57	10	30

Table 7.
Part of different combinations made.

the selling price of the fuel but also the percentage of the total dataset used in the training stage (60%). In fact, this percentage will not allow us to predict 40% of the total dataset. We will have to increase the percentage of training. In the next step, we will consider 80% of the total dataset for the training phase and 20% for testing the model. **Table 8** summarizes the different combinations [27].

The combination that minimizes the error is therefore:

- goal = 0.01;
- spread = 1;
- MN = 20; and
- DF = 30.

We can conclude that learning with 80% of the database gives increased results in comparison with the other case (learning with 60%) since the error is minimized. The output is calculated and presented in **Table 9**.

From **Table 9**, we can clearly see that the selected model can be used to model and forecast future sales in this petroleum manufacturing. As a last part, we will use the methodology of support vector machines to see that this is going to give a result.

3.1.8 Forecasting using support vector machines (SVMs)

The aim of our current work is to develop a relationship between experimental data collected from authentic sources to estimate the selling price of fuel. We are trying to apply support vector machines based on machine learning approaches because of the complex relationships between the input parameter and the output.

We prepared our database and then developed the program in Python language, which will be compiled on Spyder software.

We imported our dataset, which is the actual price of our fuel studied, created, and indexed the location of values from the database. Then, we standardized the data so that it corresponds to the learning process that will be carried out using the SVR function. In fact, we have divided our database into a learning part and another for the test. We tried two main distributions: (1) 60% of our database used in the learning phase and 40% used in the testing phase and (2) 80% of our database used in the learning phase and 20% used in the testing phase. We have kept the second distributions based on the results obtained after compiling the program. After that, we learned "Train X" and "Train Y" and executed the test to finally calculate the average of the errors and obtained the values predicted in the test phase, which are grouped in **Figure 16**.

The average error is equal to 26.882361, which represents 2.53%. The error graph is shown in **Figure 17**.

Parameters				**Relative error (%)**
Goal	**Spread**	**MN**	**DF**	
0.01	1	10	30	9.37
0.01	0.25	25	25	7.61
0.01	0.5	30	20	5.29
0.01	0.8	30	20	3.21
0.01	1	20	30	**1.95**

Table 8.
Error comparison for several combinations of parameters.

It is clear that the model chosen can be used to model and forecast future sales for this petroleum industry since the error observed (2.53%) respects the allowable margin of error set by the company at 3%. In addition, the SVR function is a useful tool, which guarantees good precision and minimizes the error compared to the ARIMA model.

Input (time)	Real value of output	Predicted value of output	% error
83	1038	1027.2	1.04046243
84	1043	1044.8	-0.1725791
85	1035	1033.4	0.15458937
86	1040	1034.3	0.54807692
87	1016	1034.7	-1.84055118
88	1015	1034.8	-1.95073892
89	1010	1034.9	-2.46534653
90	1001	1034.9	-3.38661339
91	1031	1034.9	-0.37827352
92	1033	1034.9	-0.1839303
93	1036	1034.9	0.10617761
94	1030	1034.9	-0.47572816
95	1000	1034.9	-3.49
96	1042	1034.9	0.68138196
97	1072	1034.9	3.4608209
98	1074	1034.9	3.6405959
99	1072	1034.9	3.4608209
100	1082	1034.9	4.35304991
101	1084	1034.9	4.5295203
102	1064	1034.9	2.73496241

Table 9.
Predicted value of output after using the RBF model.

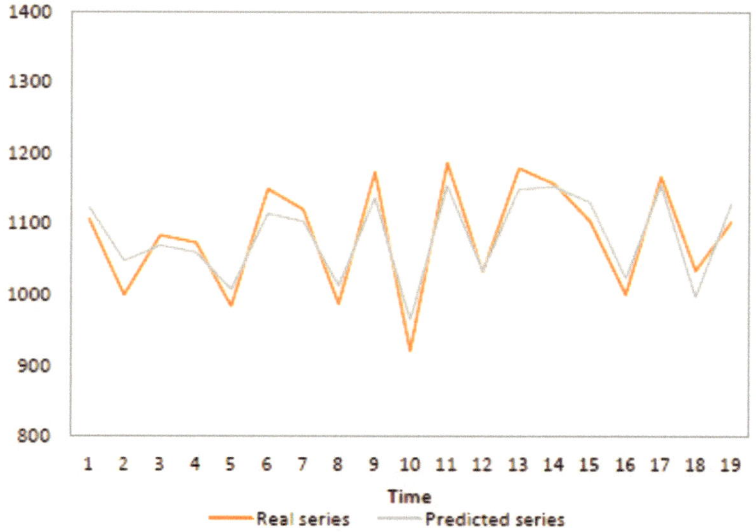

Figure 16.
Results of the SVR function.

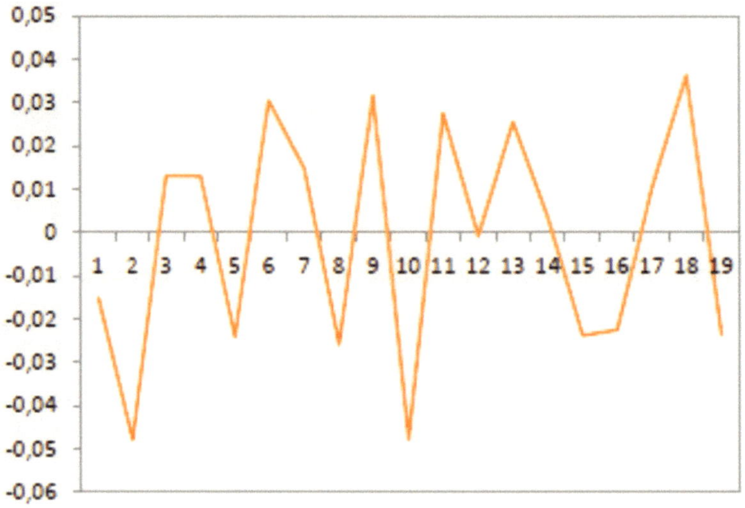

Figure 17.
Error graph.

3.2 Synthesis

In the first industrial application of this chapter, we modeled demand using ARIMA models. The model we have obtained will allow the company to forecast demand and make precise forecasts.

In the second application, we studied the selling prices of the SSP via three methodologies: ARIMA, RBF, and SVMs.

First, we developed an ARIMA model based on historical data. This study allowed us to determine the ARIMA model (1,1,1), which gives gasoline price forecasts close to the margin to reach for the first quarter of the current year with an average margin of error 2.855%. Second, we used the RBF technique to improve the modeling and forecasting of the selling price of fuel. It was found that this technique has proven its strength manifested in the error, which has been further minimized: 1.95% instead of 2.85% for the ARIMA model. Finally, we used the SVM function. The forecasts made are quite satisfactory because they respect the margin tolerated by the company. The error of the SVM function is around 2.53%.

As a summary, the SVM function has proven its strength manifesting itself in the error, which has been further minimized: 2.53% instead of 2.885% for the ARIMA model, but which remains higher than the error obtained using the RBF technique.

4. Conclusion

For most companies, forecasting is a prerequisite for effective supply chain management. Forecasting is the basis of all production management systems. The entire supply chain is based on data from forecast models.

In this chapter, we have presented the study of forecasting demand and selling prices in industrial companies. We also carried out a comparative study aimed at minimizing the error to guarantee increased forecasts.

In the first part, we modeled the future demand for a food company using ARIMA models based on the Box-Jenkins methodology. The model we have obtained will allow the company to forecast demand and make precise forecasts.

We can clearly see that the chosen model can be used to model and forecast future demand for this agribusiness, but each time we need to populate the historical data with the new data.

Second, we carried out a study, which consists in comparing the quality of the forecasts obtained in the context of forecasting selling prices. We presented the application of three different methodologies allowing us to make sales forecasts in a company operating in the petroleum sector.

We have developed an ARIMA model based on historical data. This study allowed us to determine the optimal autoregressive, moving average, and differentiation parameters in order to make predictions. We found that the ARIMA model (1,1,1) gives gasoline price forecasts close to the margin to reach for the first quarter of the current year with an average margin of error of 2.855% included within the margin of error tolerated by the company (plus or minus 3% as margin of error). In addition, the hypothesis that the residues are white Gaussian noise has always been verified.

Then, we tried forecasting selling prices via the RBF technique in order to improve the modeling and forecasting done before. To do this, we have developed an RBF network based on historical data to come up with conclusions in terms of superiority of forecast performance. Consequently, the use of this technique has proven itself and has allowed us to minimize the error, which is 1.95% versus 2.85% for the ARIMA model.

Finally, we studied the SSP selling prices via the SVM function. We prepared our database and then developed the program in Python language, which will be compiled on Spyder software. The forecasts made are quite satisfactory with regard to the constraint imposed by the company (plus or minus 3% margin of error). The error of the SVM function is around 2.53%. Consequently, the SVM function has proven its strength manifesting itself in the error, which has been further minimized: 2.53% instead of 2.855% for the ARIMA model, but which remains higher by comparing it with the error obtained if we had opted for neural networks.

Author details

Zineb Aman[1*], Latifa Ezzine[2], Yassine Erraoui[1], Younes Fakhradine El Bahi[3] and Haj El Moussami[1]

1 Mechanics and Integrated Engineering, ENSAM School, Moulay Ismail University, Meknes, Morocco

2 Modeling, Control Systems and Telecommunications, EST, Moulay Ismail University, Meknes, Morocco

3 Industrial Management and Innovation, Faculty of Science and Technology, Settat, Morocco

*Address all correspondence to: zineb.aman@gmail.com

References

[1] Lai F, Zhao X, Lee T. Selecting forecasting model parameters in material requirement planning systems. International Journal of Internet and Enterprise Management. 2006;**4**(4):331-354

[2] Fleischmann B, Meyr H, Wagner M. Supply Chain Management & Advanced Planning. Dans Berlin: Springler; 2002

[3] Stadtler H. Supply chain management and advanced planning—Basics, overview and challenges. European Journal of Operational Research. 2005;**163**(3):575-588

[4] Ghobbar AA, Friend CH. Evaluation of forecasting methods for intermittent parts demand in the field of aviation: A predictive model. Computers and Operations Research. 2003;**30**:2097-2114

[5] Toktay B. Forecasting product returns. In: Guide VDR Jr, van Wassenhove LN, editors. Business Aspects of Closed-Loop Supply Chains. Pittsburgh: Carnegie Mellon University Press; 2003

[6] Chopra S, Meindl P. Supply chain management. Strategy, planning & operation. In: Das summa summarum des management. Gabler; 2007. pp. 265-275

[7] Bozarth CB, Handfield RB. Introduction to Pperations and Supply Chain Management. 4th ed. Raleigh: North Carolina State University; 2016

[8] Wisner JD, Tan KC, Leong GK. Principles of Supply Chain Management: A Balanced Approach. Thomson South-Western: Mason; 2011

[9] Miller JJ, McCahon CS, Miller JL. Foodservice forecasting with simple time series models. Journal of Hospitality and Tourism Research. 1991;**14**:9-21

[10] Song H, Li G. Tourism demand modeling and forecasting—A review of recent research. Tourism Management. 2008;**29**:203-220

[11] Willemain TR, XSmart CN, Schwarz HF. A new approach to forecasting intermittent demand for service parts inventories. International Journal of Forecasting. 2004;**20**:375-387

[12] Taylor JW, de Menezes LM, McSharry PE. A comparison of univariate methods for forecasting electricity demand up to a day ahead. International Journal of Forecasting. 2006;**1**:1-16

[13] Weron R. Electricity price forecasting: A review of the state-of-the-art with a look into the future. International Journal of Forecasting. 2014;**30**:1030-1081

[14] Mitsutaka M, Akira I. Examination of demand forecasting by time series analysis for auto parts remanufacturing. Journal of Remanufacturing. 2015;**5**:1. DOI: 10.1186/s13243-015-0010-y

[15] Gardner ES Jr. Exponential smoothing: the state of the art—Part II. International Journal of Forecasting. 2006;**22**:637-666

[16] Liu Q, Wang H. Research on the forecast and development of China's public fiscal revenue based on ARIMA model. Theoretical Economics Letters. 2015;**5**:482-493

[17] Shen S, Shen Y. ARIMA model in the application of Shanghai and Shenzhen stock index. Applications of Mathematics. 2016;**7**:171-176

[18] Stafford J, Bodson P. L'Analyse multivariée avec SPSS. Presses de l'Universitéde Québec; 2006

[19] Bavaud F, Capel R, Crettaz de Roten F, Müller JP. Guide de l'nalyse

Statistique deDonnées avec SPSS 6. Slatkine: Genève; 1996

[20] Mélard G. Algorithm AS197: A fast algorithm for the exact likelihood of autoregressive-moving average models. Journal of the Royal Statistical Society Series C: Applied Statistics. 1984;**33**:104-114

[21] Mélard G, Roy R. Modèles de séries chronologiques avec seuils. Revue de StatistiqueAppliquée. 1988;**4**(36):5-23

[22] Mélard G. Initiation à l'analyse des séries temporelles et à la prévision. Revue MODULAD. 2006;**35**:82-129

[23] Box G, Jenkins G. Time Series Analysis: Forecasting and Control. 3rd ed. San Francisco: Holden-Day; 1994

[24] Brockwell PJ, Davis RA. Time Series: Theory and Method. Springer-Verlag; 1987

[25] Hamilton JD. Time Series Analysis. Princeton: Princeton University Press; 1994

[26] El Bahi YF, Ezzine L, El Moussami H, Aman Z. Modeling and forecasting of fuel selling price using time series approach: Case study. In: 5th International Conference on Control, Decision and Information Technologies (CoDIT). 2018

[27] Aman Z, Ezzine L, El Bahi YF, El Moussami H. Improving the modeling and forecasting of fuel selling price using the radial basis function technique: A case study. Journal of Algorithms & Computational Technology. 2019